Virtual Machining
Using CAMWorks 2016

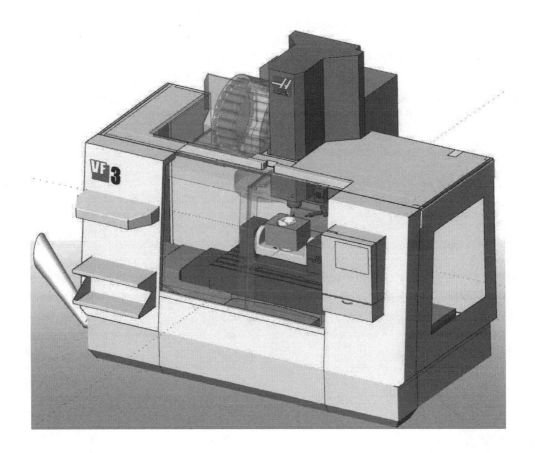

Kuang-Hua Chang, Ph.D.
School of Aerospace and Mechanical Engineering
The University of Oklahoma
Norman, OK

SDC
Publications

SDC Publications
P.O. Box 1334
Mission, KS 66222
913-262-2664
www.SDCpublications.com
Publisher: Stephen Schroff

Examination Copies
Books received as examination copies are for review purposes only and may not be made available for student use. Resale of examination copies is prohibited.

Electronic Files
Any electronic files associated with this book are licensed to the original user only. These files may not be transferred to any other party.

Trademarks
CAMWorks is a registered trademark of Geometric Americas, Inc.
SOLIDWORKS and SOLIDWORKS Simulation are registered trademarks of Dassault Systèmes Corporation.

The author and publisher of this book have used their best efforts in preparing this book. These efforts include the development, research and testing of the material presented. The author and publisher shall not be liable in any event for incidental or consequential damages with, or arising out of, the furnishing, performance, or use of the material.

ISBN-13: 978-1-63057-087-3
ISBN-10: 1-63057-087-7

Printed and bound in the United States of America.

Preface

Virtual machining is the use of simulation-based technology, in particular, computer-aided manufacturing (CAM) software, to aid engineers in defining, simulating, and visualizing machining operations for parts or assembly in a computer, or virtual, environment. By using virtual machining, the machining process can be defined and verified early in the product design stage. Some, if not all, of the less desirable design features in the context of part manufacturing, such as deep pockets, holes or fillets of different sizes, or cutting on multiple sides, can be detected and addressed while the product design is still being finalized. In addition, machining-related problems, such as undesirable surface finish, surface gouging, and a tool or tool holder colliding with stock or fixtures, can be identified and eliminated before mounting a stock on a CNC (computer numerical control) machine at shop floor. In addition, manufacturing cost, which constitutes a significant portion of the product cost, can be estimated using the machining time estimated in the virtual machining simulation.

Virtual machining allows engineers to conduct machining process planning, generate machining toolpaths, visualize and simulate machining operations, and estimate machining time. Moreover, the toolpaths generated can be converted into NC codes (also called G-codes or M-codes) to machine functional parts as well as die or mold for part production. In most cases, the toolpath is generated in a so-called CL (cutter location) data format and then converted to G-codes using respective post processors.

This book is written for the purpose of helping readers to learn the concepts and steps in conducting virtual machining using CAMWorks. CAMWorks is embedded in SOLIDWORKS as a fully integrated module and is the first SOLIDWORKS certified gold product for CAM software. CAMWorks, developed and commercialized by Geometric Americas Inc. (www.camworks.com/about), provides excellent capabilities for support of machining simulations in a virtual environment. Capabilities in CAMWorks allow users to select CNC machines and tools, extract or create machinable features, define machining operations, and simulate and visualize machining toolpaths. In addition, the machining time estimated in CAMWorks provides an important piece of information for estimating product manufacturing cost without physically manufacturing the product.

The book covers basic concepts and frequently used commands and options required for readers to advance from a novice to an intermediate level in using CAMWorks. Basic concepts and commands introduced include extracting machinable features (such as 2.5 axis features), selecting a machine and tools, defining machining parameters (such as feedrate), generating and simulating toolpaths, and post processing CL data to output G-codes for support of CNC machining. The concept and commands are introduced in a tutorial style presentation using simple but realistic examples. Both milling and turning operations are included.

One of the unique features of this book is the incorporation of the CL data verification by reviewing the G-codes generated from the toolpaths. This helps readers understand how the G-codes are generated by using the respective post processors, which is an important step and an ultimate way to confirm that the toolpaths and G-codes generated are accurate and useful.

Example files are prepared for you to go over all lessons in this book. You may download them from the publisher's website: www.sdcpublications.com.

This book is intentionally kept simple. It primarily serves the purpose of helping readers become familiar with CAMWorks in conducting virtual machining for practical applications. This is not a reference manual of CAMWorks. You may not find everything you need in this book for learning CAMWorks. But this book provides you with basic concepts and steps in using the software, as well as discussions on the G-codes generated. After going over this book, readers should develop a clear understanding in using CAMWorks for virtual machining simulations, and should be able to apply the knowledge and skills acquired to carry out machining assignments and bring machining consideration into product design in general.

This book should serve well for self-learners. A self-learner should have a basic physics and mathematics background, preferably a bachelor or associate degree in science or engineering. We assume that readers are familiar with basic manufacturing processes, especially milling and turning. Details related to basic milling, turning, and hole making can be found in textbooks, such as *Manufacturing, Engineering & Technology, 6th ed.*, by Serope Kalpakjian and Steven R. Schmid. In addition, we assume readers are familiar with G-codes. If not, we encourage you to review NC programming books, for example, *Technology of Machine Tools, seventh ed.*, by Krar, Gill, and Smid. If you are interested in understanding how the toolpaths are generated, i.e., the theory and behind-the-scene computation, you may refer to books, such as *e-Design, Computer-Aided Engineering Design*, and *Product Manufacturing and Cost Estimating using CAE/CAE*, written by the author. And certainly, we expect that you are familiar with SOLIDWORKS part and assembly modes. A self-learner should be able to complete the ten lessons of this book in about forty hours. An investment of forty hours will advance readers from a novice to an intermediate user level, a good and wise investment.

This book also serves well for class instruction. Most likely, it will be used as a supplemental reference for courses like CNC Machining, Design and Manufacturing, Computer-Aided Manufacturing, or Computer-Integrated Manufacturing. This book should cover four to five weeks of class instruction, depending on the course arrangement and the technical background of the students. Some of the exercise problems provided at the end of individual lessons may take noticeable efforts for students to complete. The author strongly encourages instructors or teaching assistants to go through those exercises before assigning them to students.

For those who desire to learn more about CAMWorks, you may find additional references on the computer where CAMWorks is installed. Several tutorial manuals provided by Geometric can be found on your computer where CAMWorks is installed:

C:\Program Files\CAMWorks2016x64\CAMWorks_VC110\Lang\English\Manuals

with example files located at:

C:\CAMWorksData\CAMWorks2016x64\Examples

Also, a few useful videos can be found on YouTube:

www.youtube.com/watch?v=JLt9HNvfjmA (general introduction)
www.youtube.com/watch?v=QnDBDUXdXa0&feature=related (2.5 axis milling)
www.youtube.com/playlist?list=PLD7F7033E2D9FB33B (3 axis milling)

and websites of technical consulting firms, such as:

Hawk Ridge Systems: www.hawkridgesys.com/products/camworks
GoEngineer: www.goengineer.com/libraries/camworks

Happy CAMWorking!!

KHC
Norman, Oklahoma
November 30, 2016

Acknowledgements

Acknowledgment is due to Mr. Stephen Schroff at SDC Publications for his encouragement and support. Without his support and the help of his staff at SDC Publications, this book would still be in its primitive stage.

Thanks are due to the undergraduate students at the University of Oklahoma (OU) for their help in testing the examples included in this book. They made numerous suggestions that improved clarity of presentation and found numerous errors that would have otherwise crept into the book. Their contributions are greatly appreciated.

I am grateful to my former students, Dr. Yunxiang Wang and Mr. Peter Staub, for their excellent efforts in carrying out the sheet metal forming project that produced results included in Lesson 10. The technical support provided by engineers and shop floor technicians, including Chris Montalbano, Mark Lucash, Jason Mann, Todd Bayles, Nate Pitcovich, and David Mason, who contributed to the success of the project is acknowledged and is highly appreciated.

About the Author

Dr. Kuang-Hua Chang is a professor for the School of Aerospace and Mechanical Engineering at the University of Oklahoma (OU), Norman, OK. He received his diploma in Mechanical Engineering from the National Taipei Institute of Technology, Taiwan, in 1980; and M.S. and Ph.D. degrees in Mechanical Engineering from the University of Iowa in 1987 and 1990, respectively. Since then, he joined the Center for Computer-Aided Design (CCAD) at Iowa as a Research Scientist and CAE Technical Area Manager. In 1997, he joined OU. He teaches mechanical design and manufacturing, in addition to conducting research in computer-aided modeling and simulation for design and manufacturing of mechanical systems.

His work has been published in 9 books and more than 150 articles in international journals and conference proceedings. He has also served as technical consultant to US industry and foreign companies, including LG-Electronics, Seagate Technology, etc. He is Associate Editor for two international journals: *Mechanics Based Design of Structures and Machines*, and *Computer-Aided Design and Applications*. He also serves on the Editorial Boards of *ISRN Mechanical Engineering*, *International Journal of Scientific Computing*, and *Journal of Software Engineering and Applications*.

About the Cover Page

The picture of the book cover was captured from a computer screen showing machining simulation that was generated for machining die and punch for support of press-forming a sheet metal part. A Kirksite block was mounted on a tilting rotary table of a HAAS mill and machining toolpath was created using CAMWorks to cut the die. In this application, the CAD model of the sheet metal part was created in SOLIDWORKS. Thereafter, sheet metal forming simulations were carried out using DynaForm (www.eta.com/inventium/dynaform) that helped explore the formability of the part and identify a set of feasible process parameters and a tooling design that would successfully form the part without tearing and excessive wrinkles. The die face was designed based on the part geometry and the punch was created by offsetting the die face with the part thickness. The finalized die surface geometry was exported from DynaForm and imported into SOLIDWORKS for tooling design. Toolpaths were generated for the die and punch machining operations using CAMWorks. A virtual HAAS mill was added to CAMWorks for machine simulation, in which collisions were detected between tool holder and the stock in the final finish operation when the tool was reaching the bottom of the pockets of the die. These issues were addressed by selecting a longer tool and a smaller tool holder in carrying out virtual machine simulation. The remedy was soon implemented on the shop floor and the collision issues were eliminated successfully. Die and punch were machined using the G-codes post processed from the toolpaths generated in CAMWorks. The machined tooling met the functional and quality requirement for part production. The sheet metal part was formed successfully on the shop floor using a 300-ton four-post press. This was a successful industrial application, where CAMWorks played a critical role in creating toolpaths to manufacture the tooling for support of sheet metal forming. More about this application and the use of CAMWorks can be found in Lesson 10: Die Machining Application.

Table of Contents

Lesson 1: Introduction to CAMWorks

1.1 Overview of the Lesson

CAMWorks, developed and commercialized by Geometric Americas Inc. (www.camworks.com/about), is a parametric, feature-based virtual machining software. By defining areas to be machined as machinable features, CAMWorks is able to apply more automation and intelligence into toolpath creation. This approach is more intuitive and follows the feature-based modeling concepts of computer-aided design (CAD) systems. Consequently, CAMWorks is fully integrated with CAD systems, such as SOLIDWORKS (and Solid Edge and CAMWorks Solids). Because of this integration, you can use the same user interface and solid models for design and later to create machining simulation. Such a tight integration completely eliminates file transfers using less-desirable standard file formats such as IGES, STEP, SAT, or Parasolid. Hence, the toolpaths generated are on the SOLIDWORKS part, not on an imported approximation. In addition, the toolpaths generated are associative with SOLIDWORKS parametric solid model. This means that if the solid model is changed, the toolpaths are changed automatically with minimal user intervention. In addition, CAMWorks is available as a standalone CAD/CAM package, with embedded CAMWorks Solids as an integrated solid modeler.

One unique feature of CAMWorks is the AFR (automatic feature recognition) technology. AFR automatically recognizes over 20 types of machinable features in solid models of native format or neutral file format; including mill features such as holes, slots, pockets and bosses; turn features such as outside and inside diameter profiles, faces, grooves and cutoffs; and wire EDM features such as die openings. This capability is complemented by interactive feature recognition (IFR) for recognizing complex multi-surface features, as well as creating contain and avoid areas.

Another powerful capability found in CAMWorks is its technology database, called TechDB™, which provides the ability to store machining strategies feature-by-feature, and then reuse these strategies to facilitate the toolpath generation. Furthermore, the TechDB™ is a self-populating database which contains information about the cutting tools and the parameters used by the operator. It also maintains information regarding the cutting tools available on the shop floor. This database within CAMWorks can be customized easily to meet the user's and the shop floor's requirements. This database helps in storing the best practices at a centralized location in support of machining operations, both in computers and on the shop floors.

We set off to learn virtual machining and explore capabilities offered by CAMWorks. The lessons and examples offered in this book are carefully designed and structured to support readers becoming efficient in using CAMWorks and competent in carrying out virtual machining simulation for general applications.

We assume that you are familiar with part and assembly modeling capabilities in SOLIDWORKS, comfortable with NC programming and G-codes, and understand the practical aspects of setting up and conducting machining operations on CNC machines on the shop floor. Therefore, this book focuses solely on illustrating virtual machining simulation using CAMWorks for toolpath and G-code generations.

Topics, such as NC part programming and transition from virtual machining to practical NC operations, can be referenced in other books mentioned in the preface of the book.

1.2   Virtual Machining

Virtual machining is a simulation-based technology that supports engineers in defining, simulating, and visualizing the machining process in a computer environment using computer-aided manufacturing (CAM) tools, such as CAMWorks. Working in a virtual environment offers advantages of ease in making adjustment, detecting error and correcting mistakes, and understanding machining operations through visualization of machining simulations. Once finalized, the toolpaths can be converted to G-codes, uploaded to a CNC machine on the shop floor to machine parts.

The overall process of using CAMWorks for conducting virtual machining consists of several steps: create design model (solid models in SOLIDWORKS part or assembly), choose NC machine and create stock, extract or identify machinable features, generate operation plan, generate toolpath, simulate toolpath, and convert toolpath to G-codes through a post processor, as illustrated in Figure 1.1. Note that before extracting machinable features, you must select an NC machine; i.e., mill, lathe, or mill-turn, choosing tool cribs, selecting a suitable post processor, and then creating a stock.

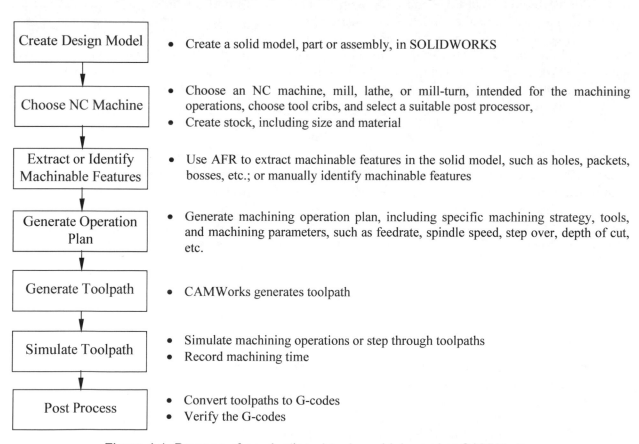

Figure 1.1 Process of conducting virtual machining using CAMWorks

The operation plan involves the NC operations to be performed on the stock, including selection of part setup origin, where G-code program zero is located. Also included is choosing tools, defining machining parameters, such as feedrate, stepover, depth of cut, etc. Note that operation plan can be automatically created by the technology database of CAMWorks as long as a machinable feature is extracted or identified manually. Users may make changes to any part of the operation plan, for instance, choosing a different tool (also called cutter, or cutting tool in the book), entering a smaller feedrate, adjusting depth of cut, etc. After an operation plan is defined, CAMWorks generates toolpath automatically. Users may simulate material removal process, step through machining toolpath, and review important machining operation information, such as machining time.

The design model (also called design part or target part in this book), which is a SOLIDWORKS part representing the perfectly finished product, is used as the basis for all machining operations. Machinable features are extracted or identified on the design model as references for each toolpath. By referencing the geometry of the design model, an associative link between the design model and the stock is established. Because of this link, when the design model is changed, all associated machining operations are updated to reflect the change.

The following example, a block with a pocket and eight holes shown in Figure 1.2, illustrates the concept of conducting virtual machining using CAMWorks. The design model consists of a base block (a boss extrude solid feature) with a pocket and eight holes that can be machined from a rectangular block (the raw stock shown in Figure 1.3) through pocket milling and hole drilling operations, respectively. A generic NC machine *Mill-in.* (3-axis mill of inch system) available in CAMWorks is chosen to carry out the machining operations. For example, toolpaths for machining the pocket (both rough and contour milling operations), as shown in Figure 1.4, can be generated referring to the part setup origin at the top left corner of the stock (see Figure 1.3). Users can step through the toolpaths, for example, the contour milling operation for cutting the pocket with tool holder turned on for display, as shown in Figure 1.5. The material removal simulation of the same toolpath can also be carried out, for example, like that of Figure 1.6.

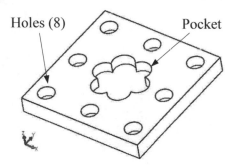

Figure 1.2 Design model in SOLIDWORKS

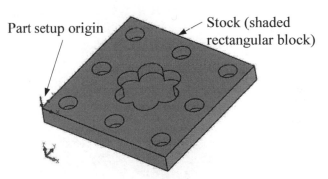

Figure 1.3 Stock enclosing the design model

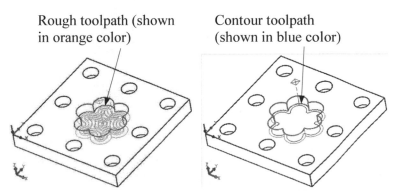

Figure 1.4 Toolpath of the pocket milling operations

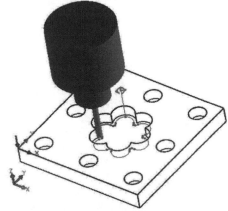

Figure 1.5 Step through toolpath

1.3   CAMWorks Machining Modules

The machining modules included in CAMWorks represent a fairly complete set of capabilities in support of virtual machining and toolpath generations. These modules include:

- 2½ axis mill: includes roughing, finishing, thread milling, face milling and single point cycles (drilling, boring, reaming, tapping) to machine prismatic features;
- 3 axis mill: includes 2.5 axis capabilities plus strategies to machine complex, contoured surfaces encountered in mold making and aerospace applications;
- 2 and 4 axis turning: includes roughing, finishing, grooving, threading, cutoff and single point cycles (drilling, boring, reaming and tapping);

Figure 1.6 The material removal simulation

- Mill-turn: includes milling and turning capabilities for multitasking machine centers;
- Multiaxis machining: 4 axis and 5 axis machining, including high-performance automotive part finishing, impellers, turbine blades, cutting tools, 5 axis trimming, and undercut machining in mold and die making;
- Wire EDM: 2.5 axis and 4 axis cutting operations automate the creation of rough, skim and tab cuts.

All the above capabilities, except wire EDM, are discussed in this book and are illustrated using simple yet practical examples. In addition, CAMWorks supports machining of multiple parts in a single setup. Parts are assembled as SOLIDWORKS assembly, which includes parts, stock, clamps, fixtures, and jig table in a virtual environment that accurately represent a physical machine setup at shop floor. A multipart machining example, as shown in Figure 1.7, with ten identical parts in an assembly will be introduced in Lesson 5.

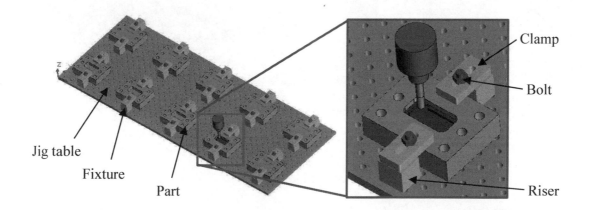

Figure 1.7 The material removal simulation of a multipart machining example

1.4   User Interface

User interface of CAMWorks is identical to that of SOLIDWORKS, as shown in Figure 1.8. SOLIDWORKS users should find it is straightforward to maneuver in CAMWorks. As shown in Figure 1.8, the user interface window of CAMWorks consists of pull-down menus, shortcut buttons, graphics area, and feature manager window. An example file, *Lesson 1 with Toolpath.SLDPRT*, is prepared for you to browse numerous capabilities and become familiar with selections, buttons, commands and options of

CAMWorks user interface. This file (and all example files of the book) is available for download at the publisher's website (www.sdcpublications.com). You may review Section 1.5 for steps to bring the example into CAMWorks.

The graphics area displays the solid or machining simulation model with which you are working. The pull-down menus provide basic solid modeling functions in SOLIDWORKS and machining functions in CAMWorks. The shortcut buttons of CAMWorks tab above the graphics area offer all the functions required to create and modify virtual machining operations in a generic order, including extract machinable features, generate operation plan, generate toolpath, simulate toolpath, step through toolpath, and post process. When you move the mouse pointer over these buttons, a short message describing the menu command will appear. Some of the frequently used buttons in CAMWorks and their functions are summarized in Table 1.1.

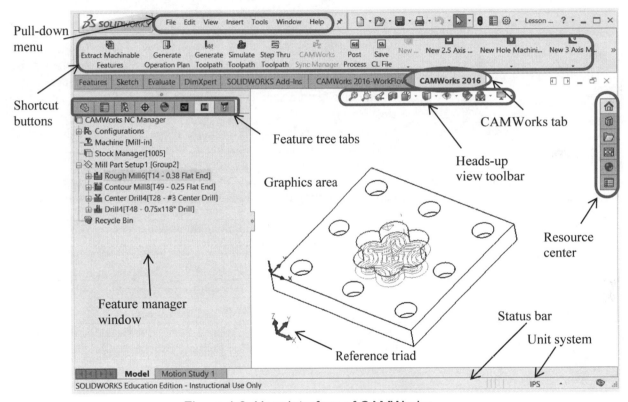

Figure 1.8 User interface of CAMWorks

There are four feature tree tabs on top of the feature manager window that are highly relevant in learning CAMWorks. The left most tab, *FeatureManager* design tree 🔧 (see Figure 1.9), sets the display to the SOLIDWORKS design tree (also called model tree or solid feature tree), which lists solid features and parts created in SOLIDWORKS part and assembly in the feature manager window.

The third tab from the right, CAMWorks feature tree 🔲 (see Figure 1.10), shifts the display to the CAMWorks feature tree, which lists machinable features extracted or identified from the solid model. The tree initially shows only the *Configurations*, *Stock Manager*, *Machine* (for example, *Mill-in* in Figure 1.10) and *Recycle Bin*. The *Machine* entity indicates the current machine as mill, turn, mill-turn, or wire EDM. You will have to select a correct machine before you begin working on a part. If you click any machinable feature, an outline view of the machinable feature appears in the part in the graphics area. For example, the sketch of the pocket appears when clicking *Irregular Pocket1* in the feature tree, as shown in

Figure 1.10. Note that a symbol ⊗ appears indicating the tool axis direction (or feed direction) of all the machinable features under the current mill part setup.

Table 1.1 The major shortcut buttons in CAMWorks

Button Symbol	Name	Function
Extract Machinable Features	Extract Machinable Features	Initiates automatic feature recognition (AFR) to automatically identify solid features that correspond to the machinable features defined in the technology database (TechDB™). The types of machinable features recognized for mill and turn are different. CAMWorks determines the types of features to recognize based on the current machine definition. The machinable features extracted are listed in the feature manager window under the CAMWorks feature tree tab [CW].
Generate Operation Plan	Generate Operation Plan	Generates operations automatically for the selected machinable features. The operations and associated parameters are based on rules defined in TechDB™. An operation contains information on how the machinable features are to be machined. The operations generated are listed in the feature manager window under the CAMWorks operation tree tab [icon].
Generate Toolpath	Generate Toolpath	Creates toolpaths for the selected operations and displays the toolpaths on the part. A toolpath is a cutting entity (line, circle, arc, etc.) created by a cutting cycle that defines tool motion.
Simulate Toolpath	Simulate Toolpath	Provides a visual verification of the machining process for the current part by simulating the tool motion and the material removal process.
Step Thru Toolpath	Step Through Toolpath	Allows you to view toolpath movements either one movement at a time, a specified number of movements or all movements.
Post Process	Post Process	Translates toolpath and operation information into G-codes for a specific machine tool controller.
Save CL File	Save CL File	Allows you to save the current operation and associated parameters in the technology database as CL (cutter location) data for future use.

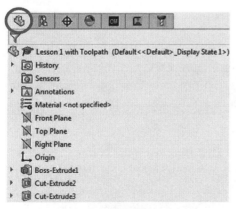

Figure 1.9 Solid features listed in the feature manager window under the *FeatureManager* design tree

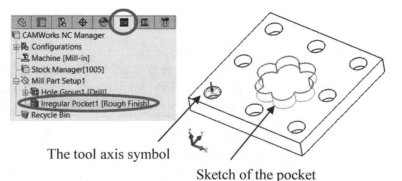

The tool axis symbol

Sketch of the pocket

Figure 1.10 Selecting a machinable feature under the CAMWorks feature tree tab

The second tab from the right, CAMWorks operation tree (shown in Figure 1.11), sets the display to the CAMWorks operation tree. After you select the *Generate Operation Plan* command, the operation tree lists operations for the corresponding machinable features. Similar to SOLIDWORKS, right clicking an operation in the operation manager tree will bring up command options that you can choose to modify or adjust the machining operation, such as feedrate, spindle speed, and so on. Clicking any operations will bring out the corresponding toolpaths in the part in the graphics area, like that of Figure 1.4.

The right most tab, CAMWorks tools tree (shown in Figure 1.12), sets the display to the CAMWorks tools tree. CAMWorks tools tree lists tools available in the tool cribs you selected for the machine.

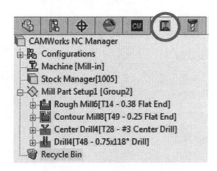

Figure 1.11 CAMWorks
operation tree tab

Figure 1.12 CAMWorks tools tree tab

1.5   Opening Lesson 1 Model and Entering CAMWorks

A machining simulation model for the simple block example shown in Figure 1.2 has been created for you. You may download all example files from the publisher's website, unzip them, and locate the model under Lesson 1 folder. Copy or move Lesson 1 folder to your hard drive.

Start SOLIDWORKS, and open model file: *Lesson 1 with toolpath.SLDPRT*. You should see a solid model file like that of Figure 1.2 appear in the graphics area.

Entering CAMWorks from SOLIDWORKS is straightforward. You may click the CAMWorks feature tree tab or operation tree tab to browse respective machining entities. You may right click any entity listed in the feature or operation tree to modify or adjust the machining model. You may also choose options under the pull-down menu *Tools > CAMWorks* to launch the same commands of those listed in Table 1.1 that support you to extract machinable features, generate operation plan, and so on.

If you do not see the CAMWorks feature tree or operation tree tab, you may have not activated the CAMWorks add-in module. To activate the CAMWorks module, choose from the pull-down menu

Tools > Add-Ins

In the *Add-Ins* dialog box shown in Figure 1.13, click *CAMWorks 2016* in both boxes (*Active Add-ins* and *Start Up*), and then click *OK*. You should see that CAMWorks 2016 tab appears above the graphics area like that of Figure 1.8 and CAMWorks tree tabs added to the top of the feature manager window.

If you do not see any of the CAMWorks tree tabs on top of the feature manager window or any of the CAMWorks buttons above the graphics area (like those of Table 1.1), you may have not set up your

CAMWorks license option properly. To check the CAMWorks license setup, choose from the pull-down menu

Help > CAMWorks 2016 > License Info

In the *CAMWorks License Info* dialog box (Figure 1.14), click all clickable boxes (or select modules as needed, for example, *3X Mill L1* for 3-axis mill level 1) to activate the module(s). Then click *OK*.

You may need to restart SOLIDWORKS to activate CAMWorks or newly selected modules. Certainly, before going over this tutorial lesson, you are encouraged to check with your system administrator to make sure CAMWorks has been properly installed on your computer.

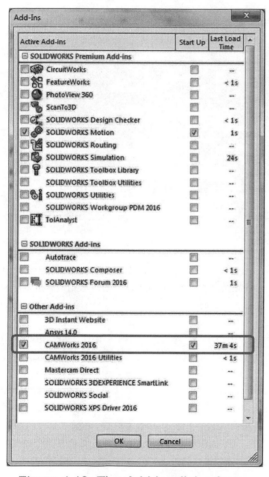

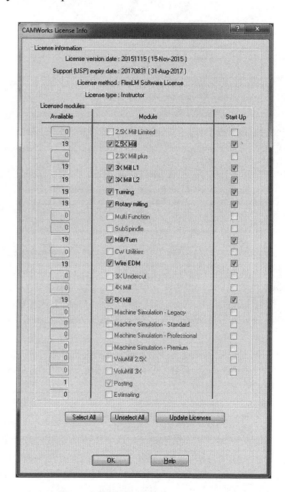

Figure 1.13 The *Add-Ins* dialog box   Figure 1.14 The *CAMWorks License Info* dialog box

Another point worth noting is that the auto save option might have been turned on in CAMWorks by default. Often, it is annoying to get interrupted by this auto save every few minutes asking if you want to save the model. You may turn this auto save option off by choosing from the pull-down menu

Tools > CAMWorks > Options

In the *Options* dialog box (Figure 1.15), select *Disable Auto Saving* under the *General* tab to turn it off. Then click *OK*.

You may click any machinable features listed under the CAMWorks feature tree tab **CW** to display the feature in the graphics area, for example, the pocket profile sketch like that of Figure 1.10 appears in the design model after clicking *Irregular Pocket1*.

You may also click any operation under CAMWorks operation tree tab to show the toolpath of the selected operation, for example, *Rough Mill* or *Contour Mill* to display toolpaths like those of Figure 1.4. You may step through toolpath of an operation by right clicking it and choosing *Step Through Toolpath* (see Figure 1.5). You may right click an operation and choose *Simulate Toolpath* to simulate a material removal process of the operation like that of Figure 1.6.

1.6   Extracting Machinable Features

Machining operations and toolpaths can be generated only on machinable features. A unique and appealing technical feature in CAMWorks is the automatic feature recognition (AFR) technology, which analyzes the solid features in the part and extracts mill features such as holes, slots, pockets and bosses; turn features such as outside and inside diameter profiles, faces, grooves and cutoffs, and wire EDM features such as die openings. The AFR technology helps in reducing the time spent by the designer to feed in data related to creating machining simulation.

Figure 1.15   The *Options* dialog box

A set of machinable features for milling and turning operations that can be extracted by AFR can be found in Appendix A. The associated machining strategies of individual machinable features are summarized in Appendix B.

The *Extract Machinable Features* command initiates AFR. Depending on the complexity of the part, AFR can save considerable time in extracting 2.5 axis features, such as holes, pockets, slots, bosses, etc., either prismatic (with vertical walls) or tapered.

AFR cannot recognize every single feature on complex parts and does not recognize features beyond 2.5 axis. To machine these areas, you need to define machinable features interactively using the interactive feature recognition (IFR) wizard. For example, you may define a *Multi Surface* feature interactively by selecting faces to be cut and faces to avoid in the design model. More on this topic will be discussed in this book, for example, in Lesson 2: Simple Plate, and Lesson 4: Freeform Surface.

In addition to AFR and IFR, CAMWorks provides local feature recognition (LFR) capability for extracting machinable features. Local feature recognition is a semi-automatic method to define machinable features based on face selection. You may select one or more faces on the part in the graphics area and use LFR to extract machinable features.

1.7   Technology Database

CAMWorks technology database, TechDB™, is a self-populating database which contains all the information about the machine, cutting tools and the parameters used by the operator, rules of repetitive NC operations (called strategy in CAMWorks) for the respective machinable features. This database within CAMWorks can be customized easily to meet the user's and the shop floor's requirements. The database can be utilized for all the manufacturing processes viz. milling, turning, mill-turn and wire EDM. This database helps the best practices at a centralized location in the tool room; thus, it eliminates the non-uniformity in practicing virtual and physical machining operations.

Using a set of knowledge-based rules, CAMWorks analyzes the machinable features to determine the machining process plan for a part. In this approach, features are classified according to the number of possible tool approaching directions that can be used to machine them. The knowledge-based rules are applied to assure that you get desired cutting operations. The rules that determine machining operations for a respective machinable feature can be found in Appendix B, for both milling and turning.

The technology database is shipped with data that is considered generally applicable to most machining environments. Data and information stored in the database can be added, modified, or deleted to meet the user's specific needs in practice. More about accessing and modifying the database is discussed in Lesson 10: Die Machining Application.

1.8   CAMWorks Machine Simulation

CAMWorks Machine Simulation offers a realistic machine setting in a virtual environment, in which machining simulation may be conducted on a virtual replicate of the physical machine. Solid models of the machine, tilt rotary table, fixtures, tool, tool holder, stock and part are assembled to realistically represent a physical NC machine setting. In addition to simulating machining operations, the Machine Simulation capability provides tool collision detection in a more realistic setting.

Although CAMWorks Machine Simulation capability is offered as a separate licensed module, legacy machines, including a mill and a mill-turn, come with CAMWorks. For example, Figure 1.16 shows machining simulation using a legacy mill, *Mill_Tutorial*, in which the tool, tool holder, tilt table, rotary table, and a stock together with a machine coordinate system XYZ are displayed (see Figure 1.16). To the right, the machining operations and the corresponding G-codes are listed in the upper and lower areas of *Move List*, respectively. On top, a default virtual machine (or simulator), *Mill_Tutorial*, is selected. Below are buttons that control the machining simulation run. More about the Machine Simulation capability and commands can be found in Lesson 7: Multiaxis Milling and Machine Simulation.

More importantly, this capability allows you to install a virtual CNC machine that is a virtual replicate of a physcial machine available at your machine shop into your computer to carry out machining simulation. For example, a HAAS mill is added to CAMWorks as a virtual CNC machine for support of machine simulation, as shown in Figure 1.17. We will discuss this capability in Lesson 10: Die Machining Application.

1.9   Tutorial Examples

In addition to the example of this lesson, nine machining examples are included in this book: seven milling and two turning. All nine lessons illustrate step-by-step details of creating machining operations and simulating toolpath capabilities in CAMWorks. We start in Lesson 2 with a simple plate example, which provides you with a brief introduction to CAMWorks 2016, and offers a quick run-through for creating a contour mill operation using a 3-axis mill.

NC machine
Tool holder
Tool
Tilt table
Rotary table
Stock

NC operations
G-codes

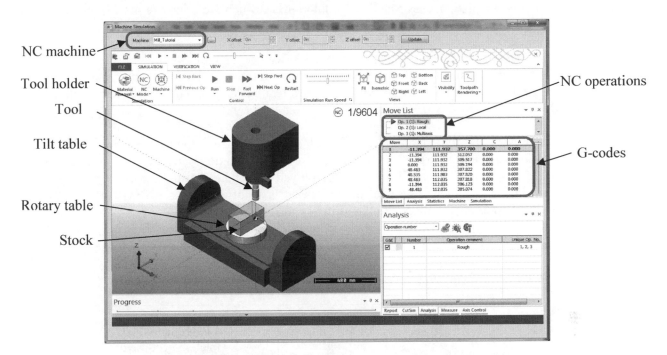

Figure 1.16 The *Machine Simulation* window of *Mill_Tutorial*

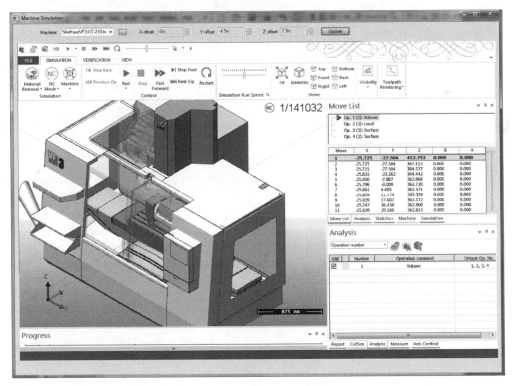

Figure 1.17 The *Machine Simulation* window of HAAS mill

Lessons 3 through 7 focus on milling operations. We include examples of machining 2.5 axis features using 3-axis mill in Lesson 3, machining a freeform surface of a solid feature in Lesson 4, machining a set of identical parts in an assembly in Lesson 5, machining features on multiple planes using a 3-axis mill with a rotary table in Lesson 6, and machining a cylindrical surface using multiaxis milling operation in

Lesson 7. In Lesson 7 we also discuss Machine Simulation, in which we bring the machining simulation into a legacy mill that comes with CAMWorks software.

Lessons 8 and 9 focus on turning operations. In Lesson 8, we use a simple stepped bar example to learn basic capabilities in simulating turning operations and understanding G-codes generated by CAMWorks. In Lesson 9, we machine a similar example with more turn features to gain a broader understanding of the turning capabilities offered by CAMWorks.

In Lesson 10, we present an industrial application that involves die machining for sheet metal forming. In this application, tooling manufacturing for sheet metal forming, including punch and die, was carried out mainly by using a HAAS mill. CAMWorks was employed to conduct virtual machining and toolpath generation for the die and punch. The goal of the lesson is to offer readers a flavor of the role that CAMWorks plays in a practical tooling manufacturing application.

All examples and topics to be discussed in individual lessons are summarized in Table 1.2.

Table 1.2 Examples employed and topics to be discussed in this book

Lesson	Example	Machining Model	Problem Type	Things to Learn
2	Simple Plate		3-axis contour milling	1. A brief introduction to CAMWorks 2016 2. A complete procedure of using CAMWorks to create a milling operation from the beginning all the way to the post process that generates G-codes 3. Extract machinable feature using interactive feature recognition (IFR) 4. Review and verify G-code generated
3	2.5 Axis Features		3-axis pocket milling and hole drilling	1. Extract 2.5 axis machinable features using automatic feature recognition (AFR) 2. Identify machinable feature for face milling operation interactively 3. Define part setup origin for G-code generation 4. Adjust machining parameters to regenerate toolpath 5. Review and verify G-code generated
4	Freeform Surface		Machine a freeform surface using 3-axis mill	1. Create a multi surface machinable feature and select avoid surface feature to restrain area of toolpath generation 2. Creating *Area Clearance* (rough cut) and *Pattern Project* (finish cut) operations 3. Converting an area clearance operation to local milling operation 4. Creating section views for a closer look at the freeform surface in material removal simulation
5	Multipart Machining		Machine a set of identical parts in an assembly	1. Create instances of part for machining operations 2. Define stocks for individual instances 3. Select components in the assembly for the tools to avoid 4. Review and verify G-code generated

Lesson	Example	Machining Model	Problem Type	Things to Learn
6	Multiplane Machining		Machine features on multiple planes using a 3-axis mill with a rotary table	1. Cut parts with machinable features on multiple planes 2. Set a rotation axis of the rotary table as the 4th axis 3. Select components to be included in material removal simulation 4. Rotate tool vs. rotate stock in material removal simulation 5. Review and verify G-code generated
7	Multiaxis Surface Machining		Multiaxis milling operations and Machine Simulation	1. Machine a cylindrical surface extruded by a Bézier curve using 5-axis mill 2. Create volume milling, local milling, and multiaxis surface milling operations 3. Identify tool gouging and regenerate toolpath to avoid gouging 4. Use Machine Simulation to simulate multiaxis machining operations in a setup with a tilt rotary table
8	Turning a Stepped Bar		Basic turning operations using 2-axis lathe	1. A complete procedure in using CAMWorks to create a turning simulation from the beginning all the way to the post process that generates G-codes 2. Create machinable feature using interactive feature recognition 3. Review and verify G-code generated
9	Turning a Stub Shaft		Advanced turning operations	1. Extract machinable features for turning, including face, groove, thread, and holes at both ends using AFR 2. Turn thread feature, and review and verify G-code generated 3. Choose mill-turn to machine the side cut features and a cross hole in the stub shaft
10	Die Machining Application		Practical application	1. Introduce an industrial application that involves die machining for sheet metal forming 2. Add a HAAS mill to CAMWorks 3. Customize the technology database 4. Use the added HAAS mill to support machine simulation 5. Add a HAAS post processor to CAMWorks for G-code generation 6. Collision detection in Machine Simulation

Lesson 2:   A Quick Run-Through

2.1   Overview of the Lesson

The purpose of this lesson is to provide you with a quick start in using CAMWorks. You will learn a complete procedure in using CAMWorks to create a machining simulation from the beginning all the way to the post process that generates G-code for CNC machining. We will use a 3-axis mill to machine a simple plate by creating a contour milling operation (often called profile milling in other CAM software) that cuts the profile boundary of the part, as shown in Figure 2.1. This lesson is intentionally made simple for new CAMWorks users. We stay with default options and parameters for most of the selections.

Figure 2.1 The material removal simulation of the simple plate example

We follow the general steps shown in Figure 1.1 of Lesson 1 to create machining simulation and generating toolpath using CAMWorks. In addition, we will go over a post process to generate G-code for this contour milling operation.

After completing this lesson, you should be able to carry out machining simulation and toolpath generation for similar problems following the same procedure. This lesson should also prepare you to go over the remaining lessons of the book.

2.2   The Simple Plate Example

The L-shape plate has an bounding box of size 4in.×4in.×0.5in., as shown in Figure 2.2, with a fillet of 0.375in. The unit system chosen is IPS (inch, pound, second). There is one solid feature created in the design model, *Boss-Extrude1*, listed in the *FeatureManager* design tree 🔲 (or simply model free or feature tree) shown in Figure 2.3. In addition, a fixture coordinate system (*Coordinate System1*) is defined at the top left corner of the part, which defines the "home point" or main zero position on the machine. When you open the solid model *Simple Plate.SLDPRT*, you should see the solid feature and the coordinate system listed in the feature tree like that of Figure 2.3.

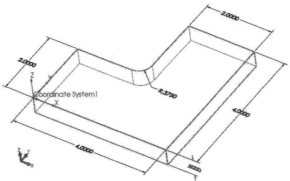

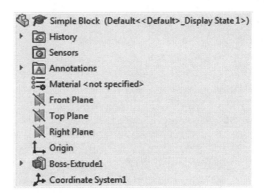

Figure 2.2 Dimensions of the simple plate model   Figure 2.3 Entities listed in the feature tree

A stock of retangular block with a size 4.5in.×4.5in.×0.5in., made of low carbon alloy steel (1005), as shown in Figure 2.4, is chosen for the machining operation. Note that a part setup origin is defined at the top left vertex of the stock, which defines the G-code program zero location.

We will create one contour milling operation, which cuts along the profile boundary of the part using a 1-in. flat-end mill. We will use the interactive feature recognition (IFR) capability to identify the machinable feature and follow the recommendations of the technology database (TechDB™) for choosing machining operation and parameters, such as feedrate and spindle speed. We will make a few changes to obtain a toolpath of the contour mill operation like that of Figure 2.5, in which the cutter moves along the profile boundary of the part in two passes. This is because the machining depth (or depth of cut), 0.3in., chosen was less than the thickness of the stock (0.5in.).

2.3   Using CAMWorks

Open SOLIDWORKS Part

Open the part file (filename: *Simple Plate.SLDPRT*) downloaded from the publisher's website. This solid model, as shown in Figure 2.2, consists of one solid feature and a coordinate system among other entities. As soon as you open the model, you may want to check the unit system chosen.

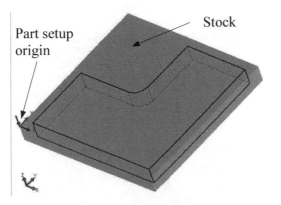

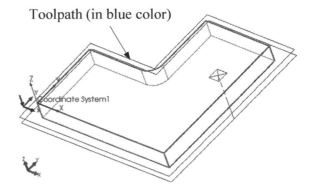

Figure 2.4 Stock with a part setup origin at its top left vertex

Figure 2.5 The toolpath of the contour milling operation

You may select from the pull-down menu *Tools > Options*. In the *Document Properties* dialog box (Figure 2.6), select the *Documents Properties* tab and select *Units*. Select *IPS (inch, pound, second)*. We will stay with this unit system for this lesson. Also, it is a good practice to increase the decimals from the default 2 to 3 or 4 digits since some machining parameters defined are down to a thousandth of an inch in CAMWorks. To increase the decimals to 4 digits, you may click the dropdown button to the right of the cell in the *Length* row under the *Decimals* column. Pull down the selection and choose *.1234* for 4 digits. Click *OK* to accept the change.

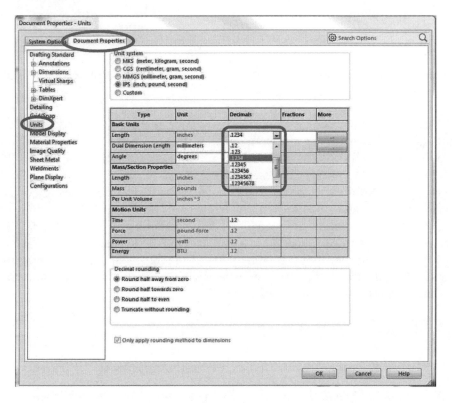

Figure 2.6 The *Document Properties* dialog box

Enter CAMWorks

At the top of the feature tree, you should see two important tabs, CAMWorks feature tree [CW] and CAMWorks operation tree [K], as shown in Figure 2.7. You may click these two tabs to review CAMWorks machinable features and machining operations, respectively. You should also see CAMWorks short-cut buttons like those of Table 1.1 above the graphics area (after clicking the *CAMWorks 2016* tab).

We will manually create a machinable feature (instead of clicking the *Extract Machinable Features* button), and use the four buttons next to it, *Generate Operation Plan* , *Generate Toolpath* , *Simulate Toolpath* , and *Step Through Toolpath* , to create and review the toolpath.

CAMWorks   CAMWorks
feature tree tab   operation tree tab

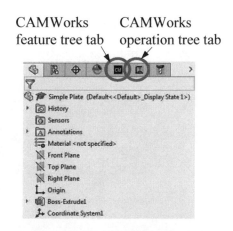

Figure 2.7 The CAMWorks feature tree and CAMWorks operation tree tabs

Before going through these steps, we will first select a NC machine and define a stock to be employed for machining this part.

Select NC Machine

Click the CAMWorks feature tree tab . A default mill *Mill-in*, which is a 3-axis mill of inch system, is listed in the feature manager window (Figure 2.8). This is the NC machine we want to use. We right click *Mill-in* and choose *Edit Definition*.

Figure 2.8 The NC machine *Mill-in* chosen by default

In the *Machine* dialog box (Figure 2.9), *Mill-in* is selected under *Machine* tab, in which you may find a few basic machine parameters of the mill. You may choose a different machine from those listed in the box under *Available machines* if needed.

We will not change the machine selection. We will choose *Tool Crib*, *Post Processor*, and *Setup* tabs to review or adjust the machine definition.

Choose the *Tool Crib* tab and select *Crib 1* under *Available tool cribs* (Figure 2.10), and then click *Select*. We will use the tools available in *Crib 1* for this lesson. Choose the *Post Processor* tab, a post processor called *M3AXIS-TUTORIAL* is selected (Figure 2.11). This is a generic post processor of 3-axis mill that comes with CAMWorks.

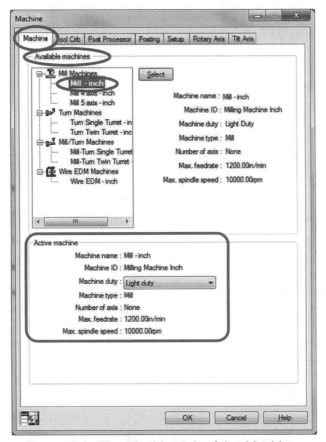

Figure 2.9 The *Machine* tab of the *Machine* dialog box

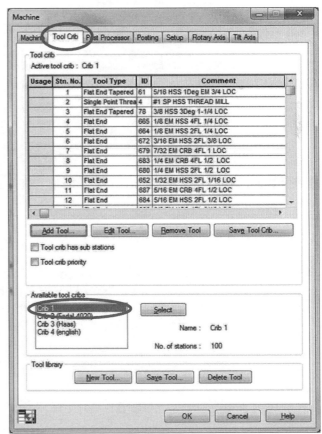

Figure 2.10 The *Tool Crib* tab of the *Machine* dialog box

There are other post processors that come with CAMWorks, which are located in *C:\CAMWorks Data\CAMWorks2016\Posts*. Note that in practice you will have to identify a suitable post processor that produces G-codes compataible with the CNC machines at shop floor. We will stay with *M3AXIS-TUTORIAL* for this lesson.

Then, we click the *Setup* tab, and choose *Coordinate System1* under *Fixture Coordinate system* (Figure 2.12). Click *OK* to accept the selections and close the dialog box.

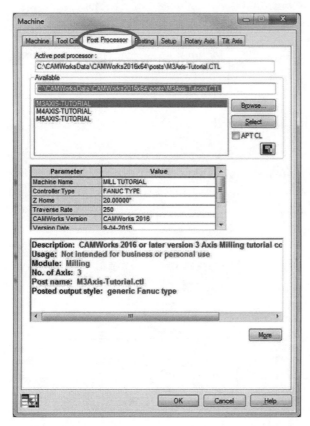

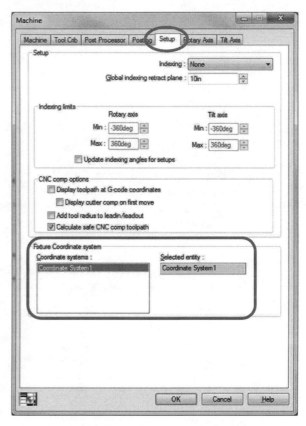

| Figure 2.11 The *Post Processor* tab of the *Machine* dialog box | Figure 2.12 The *Setup* tab of the *Machine* dialog box |

Create Stock

From CAMWorks feature tree , right click *Stock Manager* and choose *Edit Definition*. The *Stock Manager* dialog box appears (Figure 2.13), in which a default stock size appears at the bottom of the dialog box, which is the size of the bounding box of the part. The default stock material is *Steel 1005*. We will increase the length and width of the stock by 0.25in. on both sides (enter 0.25 for *X+*, *X-*, *Y+*, and *Y-*, as shown in Figure 2.13). The stock size becomes X:4.5, Y:4.5, Z:0.5, as circled in the *Stock Manager* dialog box (Figure 2.13). Accept the revised stock by clicking the checkmark ✔ at the top left corner. The rectangular stock should appear in the graphics area similar to that of Figure 2.4.

Mill Part Setup and Machinable Feature

Since machinable feature of the contour milling is not extracted by CAMWorks automatically, we will create one manually. We will first create a mill part setup and then insert a 2.5 axis feature.

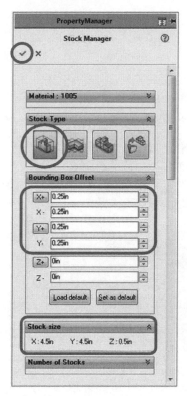

Figure 2.11 The *Stock Manager* dialog box

Figure 2.12 The *Mill Setup* dialog box

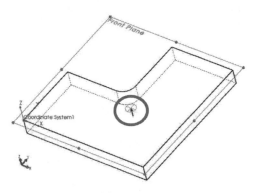

Figure 2.13 The *Front Plane* selected

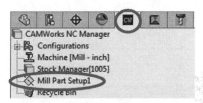

Figure 2.14 *Mill Part Setup1* listed in the feature tree

From the CAMWorks feature tree [CW], right click *Stock Manager* and choose *New Mill Part Setup*. The *Mill Setup* dialog box appears (Figure 2.14) with the *Front Plane* selected under *Entity*. In the graphics area (similar to that of Figure 2.15), a tool axis symbol ⌀ with arrow facing upward appears. This symbol indicates the tool axis or the feed direction that must be reversed (pointing downward).

Click the *Reverse Selected Entity* button [↗] under *Entity* to reverse the direction. Make sure that the arrow points in a downward direction. Click the checkmark ✓ to accept the definition. A *Mill Part Setup1* is now listed in the feature tree, as shown in Figure 2.16.

Now we define a machinable deature. From CAMWorks feature tree [CW], right click *Mill Part Setup1* just created and choose *New 2.5 Axis Feature*.

In the *2.5 Axis Feature* dialog box (Figure 2.17), choose *Boss* for *Type*, and select *Sketch1* under *Available Sketches*, then click the *Next* button ⊙ (circled in Figure 2.17) to define end condition.

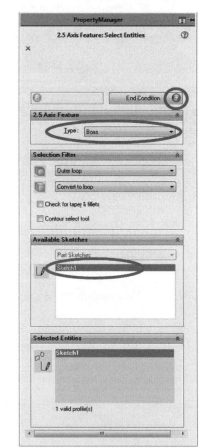

Figure 2.15 The *2.5 Axis Feature* dialog box

Note that *Sketch1* defines the outer profile of the plate. Choose *Coarse* for *Strategy* (Figure 2.18), and enter *0.5* for the depth dimension under *End condition – Direction 1*, then click the checkmark to accept the definition.

As soon as a strategy is selected, CAMWorks displays a set of machining parameters generated by the TechDB™, as shown in Figure 2.19, in which a 2in. diameter flat-end cutter is selected. In addition, tool material, XY and Z feedrates, and machining depth (0.5in.) are determined among others.

Note that it is apparent that a 2in. diameter tool will not be able to cut the fillet of 0.375in. in the part. A tool of diameter 0.75in. or less will be adequate. We will need to choose a suitable tool shortly.

A *Irregular Boss1* entity is now listed in the CAMWorks feature tree in magenta color (Figure 2.20).

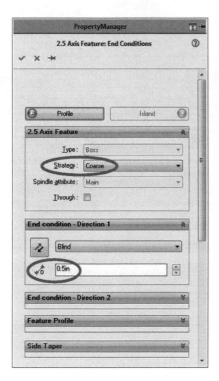

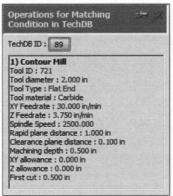

Figure 2.17

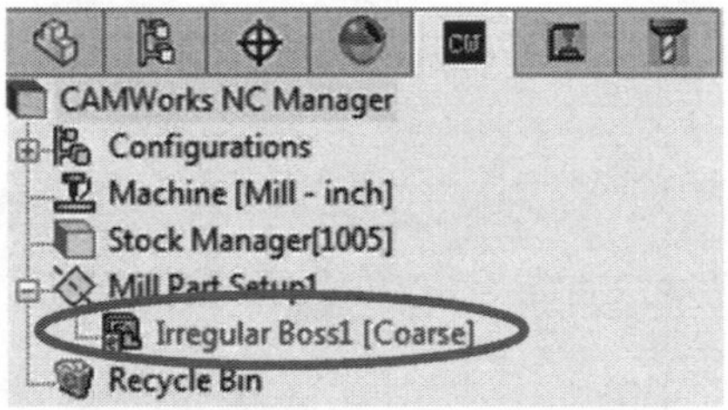

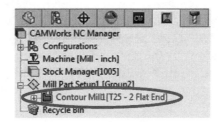

Figure 2.18 Machinable feature
added to the feature tree

Figure 2.16 Defining end
conditions

Figure 2.19 Machinable operation
added to the operation tree

Generate Operation Plan and Toolpath

Right click *Irregular Boss1* and choose *Generate Operation Plan* (or click the *Generate Operation Plan* button above the graphics area).

A new entity, *Contour Mill1*, is listed in CAMWorks operation tree (see Figure 2.21). Right click *Contour Mill1* and choose *Generate Toolpath* (or click the *Generate Toolpath* button above the graphics area).

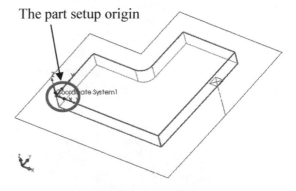

The part setup origin

Figure 2.20 The toolpath generated

A contour milling toolpath will be generated like that shown in Figure 2.22. Note that the flat-end cutter cuts along the part profile in one pass since the machining depth (or depth of cut) was chosen as 0.5in. by CAMWorks.

Define Part Setup Origin

The toolpath generated, as shown in Figure 2.22, assumes a part setup origin (see the symbol $\overset{\curvearrowleft}{\kappa}$, circled in Figure 2.22) located at top left corner of the design part, coinciding with the coordinate system, *Coordinate System1*.

As mentioned before, the part setup origin defines the program zero for the G-code to be generated after the toolpath is finalized. In practice, a part setup origin defined on the design part is less desirable. It is more practical to set the origin at a corner of the stock, for example, its top left corner, which is accessible and easy to set up on the mill.

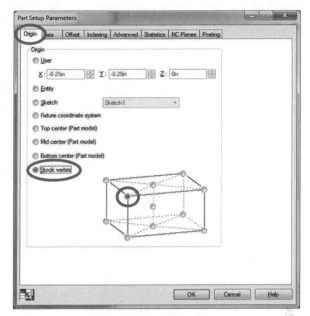

Figure 2.21 The *Origin* tab of the *Part Setup Parameters* dialog box

We right click *Mill Part Setup* under the CAMWorks operation tree tab ![icon] and choose *Edit Definition*. In the *Part Setup Parameters* dialog box (Figure 2.23), select the *Origin* tab, choose *Stock vertex* and pick the vertex at the front left corner of the top face of the sample stock (circled in Figure 2.23). The part setup origin is moved to the front left corner of the top face of the stock like that of Figure 2.4. Click *OK* to accept the change.

Click the *Yes* button in the CAMWorks warning box: *The origin or machining direction or advanced parameters has changes, toolpaths need to be recalculated. Regenerate toolpaths now?* The toolpath will be regenerated referring to where the setup origin was relocated.

Modify the Toolpath

Next we replace the tool with a 0.75in. flat-end cutter, and reduce the machining depth to 0.3in.

Under the CAMWorks operation tree tab ![icon], right click *Contour Mill1* and choose *Edit Definition*. In the *Operation Parameters* dialog box, choose *Tool* tab; the 2in. flat-end cutter is shown (Figure 2.24). Click the *Tool Crib* tab, select the 21ˢᵗ tool (Type: *FLAT END*, ID: *709*, Comment: *3/4 EM HSS 4FL 1LOC*), click the *Select* button to select the tool (Figure 2.25).

Click *Yes* to the question in the warning box: *Do you want to replace the corresponding holder also?*

Choose the *Contour* tab of the *Operation Parameters* dialog box. In the *Depth* parameters group, click the percentage button ![icon] to deselect it for both *First cut amt.* and *Max cut amt.* Enter *0.3in* for both *First cut amt.* and *Max cut amt.*, as shown in Figure 2.26, since a smaller tool has been chosen.

Click the *F/S* tab to review machining parameters (Figure 2.27); for example, the *XY feedrate* is chosen as *106.7in./min.* by the techology database of CAMWorks. Click *OK* to accept the changes.

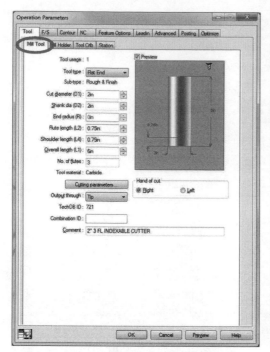

Figure 2.22 The *Mill Tool* tab of the
Operation Parameters dialog box

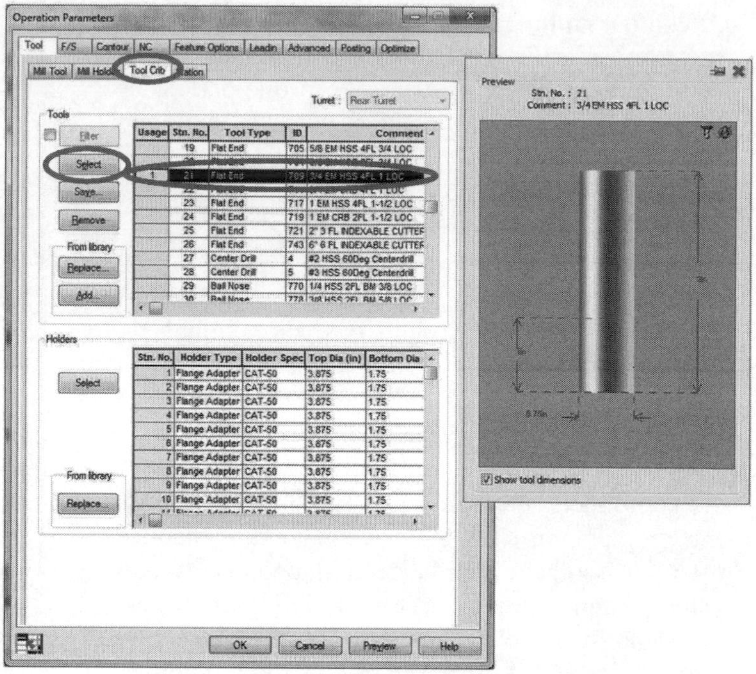

Figure 2.23 Selecting a 0.75in. flat-end cutter
under the *Tool Crib* tab

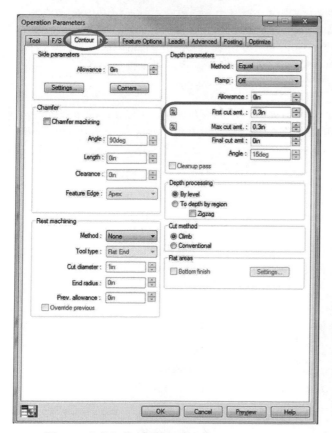

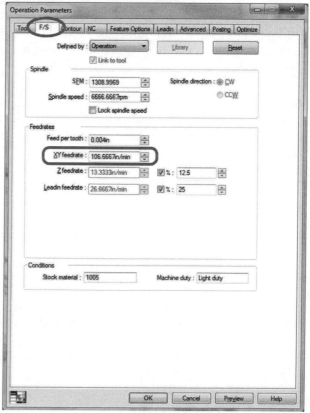

Figure 2.24 Defining depth parameters
under the *Contour* tab

Figure 2.25 The *XY feedrate* shown under
the *F/S* tab

Right click *Contour Mill1* entity and choose *Generate Toolpath*. Profile milling toolpath will be generated like that shown in Figure 2.5 with two passes.

Simulate Toolpath

Right click *Contour Mill1* entity and choose *Simulate Toolpath* (or click the *Simulate Toolpath* button  above the graphics area). The *Toolpath Simulation* tool box appears (Figure 2.28).

Click the *Run* button ▶ circled in Figure 2.28 to simulate the toolpath, or carry out the material removal simulation. The machining simulation of the contour milling operation will appear in the graphics area, similar to that of Figure 2.1.

Figure 2.26 Play the machining simulation by clicking the *Run* button

Step Through Toolpath

You may step through the toolpath to learn more about the individual machining steps. Right click *Contour Mill1* entity and choose *Step Thru Toolpath* (or click the *Step Thru Toolpath* button above the graphics area). The *Step Through Toolpath* dialog box appears (Figure 2.29). Under *Information*, CAMWorks shows the tool movement from the current to the next steps (in X, Y, and Z coordinates), the feedrate and spindle speed, among others.

Click *Step* button ▷| at the center of the tool box (circled in Figure 2.29) to step through the toolpath. You may want to turn on *Show toolpath points*, and *Tool Holder Shaded Display* (circled in Figure 2.29) to see the toolpath displayed on the part similar to that of Figure 2.30.

2.4   The Post Process and G-Code

Now we have completed the contour milling operation. We learn how to convert the toolpath to G-code and CL data next.

Right click *Contour Mill1* entity and choose *Post Process* (or click the *Post Process* button above the graphics area). In the *Post Output File* dialog box (Figure 2.31), choose a proper file folder, enter a file name (default name is *Simple Plate.txt*), use the default type (*M3Axis-Tutorial*), and click *Save*. The *Post Process Output* dialog box appears (Figure 2.32).

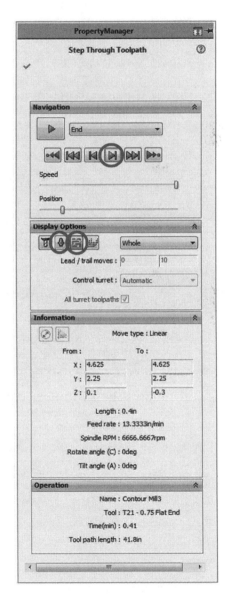

Figure 2.27 The *Step Thru Toolpath* dialog box

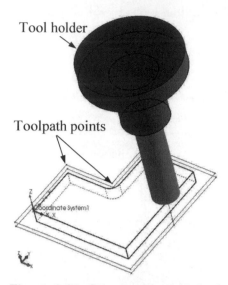

Tool holder

Toolpath points

Figure 2.28 Stepping through toolpath

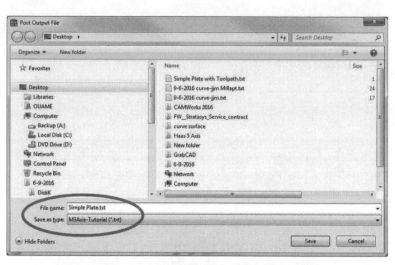

Figure 2.29 The *Post Output File* dialog box

Figure 2.30 The *Post Process Output* dialog box

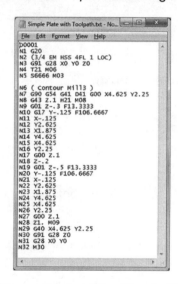

Figure 2.31 The contents of the *Simple Plate.txt* file

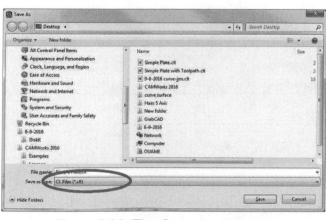

Figure 2.32 The *Save As* dialog box

```
UNITS/ INCHES
BOX/ 2.000000,2.000000,-0.250000,4.000000,4.000000,0.500000
CUTTER/
0.750000,0.000000,0.375000,0.000000,0.000000,0.000000,3.000000,
TLAXIS/ 0.000000,0.000000,1.000000
OPFEATSTART/Contour Mill3-Irregular Boss1
ROTABL/ 0.000000,AAXIS,TABLE,CCLW
ROTABL/ -0.000000,BAXIS,TABLE,CCLW
ROTABL/ 0.000000,CAXIS,TABLE,CCLW
CUTCOM/ ON
CUTCOM/ LEFT
RAPID/
GOTO/ 4.625000,2.250000,1.000000
RAPID/
GOTO/ 4.625000,2.250000,0.100000
FEDRAT/ IPM,13.333333
GOTO/ 4.625000,2.250000,-0.300000
FEDRAT/ IPM,106.666667
GOTO/ 4.625000,-0.125000,-0.300000
GOTO/ -0.125000,-0.125000,-0.300000
```

Figure 2.33 The partial contents of the
CL file (*Simple Plate.clt*)

In the *Post Process Output* dialog box, click the *Play* button (circled in Figure 2.32) to create a G-code file (.txt file).

Open the G-code file (filename: *Simple Plate.txt*) from the folder using *Word* or *Word Pad* (see the file contents shown in Figure 2.33 with command explanations provided in Column B of Table 2.1).

Table 2.1 Contents of the .txt file (NC codes) with explanations

NC Commands (Column A)	Explanations (Column B)
O0001	Main program, program #0001
N1 G20	G20: Select Inches
N2 (3/4 EM HSS 4FL 1 LOC)	Comments in parentheses
N3 G91 G28 X0 Y0 Z0	G91: Incremental programming, G28: Return To Reference Point, move cutter to X0Y0Z0
N4 T21 M06	M06: Tool Change, Load tool #T21
N5 S6666 M03	M03: Spindle forward at speed 6666rpm
N6 (Contour Mill3)	Comments
N7 G90 G54 G41 D41 G00 X4.625 Y2.25	G90: Absolute programming, G54: Select Work Coordinate System 1, G41: 2D Cutter Compensation Left, D41: tool diameter registered at #41, G00: move cutter to Point A rapidly (see Figure 2.37 for the toolpath points)
N8 G43 Z.1 H21 M08	G43: tool length compensation with length registered for the cutter #21, M08: coolant on
N9 G01 Z-.3 F13.3333	G01: cutting, plunge the cutter 0.3in. into the stock at feedrate 13.333 in/min
N10 G17 Y-.125 F106.6667	G17: cutting on the XY plane, move the cutter to Point B at feedrate 106.667 in/min
N11 X-.125	Point C
N12 Y2.625	Point D
N13 X1.875	Point E
N14 Y4.625	Point F
N15 X4.625	Point G
N16 Y2.25	Point A
N17 G00 Z.1	Retract to 0.1in. in the Z-direction, rapid
N18 Z-.2	Move to -0.2 in the Z-direction, rapid
N19 G01 Z-.5 F13.3333	G01: cutting, plunge the cutter 0.5in. into the stock at feedrate 13.333 in/min for the second pass
N20 Y-.125 F106.6667	Move the cutter to Point B at feedrate 106.667 in/min
N21 X-.125	Point C
N22 Y2.625	Point D
N23 X1.875	Point E
N24 Y4.625	Point F
N25 X4.625	Point G
N26 Y2.25	Point A
N27 G00 Z.1	Retract to 0.1in. in the Z-direction, rapid
N28 Z1. M09	Retract to 1.0 in. in the Z-direction, rapid, M09: Coolant Off
N29 G40 X4.625 Y2.25	G40: 2D Cutter Compensation Off, move cutter to Point A
N30 G91 G28 Z0	G91: Incremental programming, G28: Return To Reference Point, move cutter to Z0
N31 G28 X0 Y0	G28: Return To Reference Point, move cutter to X0Y0
N32 M30	Program End and Rewind

Note that the NC word G41 in block N7 turns on tool radius compensation from the left, which moves the tool to the left of the program path by the amount of the tool radius. Since the cutter locations of the G-code specify the actual locations of the tool already (such as Points A, B, C, etc. shown in Figure 2.37), tool radius compensation is not necessary. The NC word G41 should have been G40 that cancels the compensation at the beginning of the NC program. This is an error created by the post processor, *M3AXIS-TUTORIAL*, which must be corrected.

After correcting the error (and perhaps a few more, depending on the controller of your NC machine), the .txt file can be uploaded to a CNC mill, for example a HAAS mill, to machine the part. In general, the G-code converted is ready to be loaded to a mill. However, a few minor changes may be needed (or errors found in this example to be corrected), depending on the post processor employed in the G-code output.

Right click *Contour Mill1* entity and choose *Save CL File* (or click the *Save CL File* button ▣ above the graphics area). The *Save As* dialog box appears (Figure 2.34).

In the *Save As* dialog box (Figure 2.34), choose a proper file folder, enter a file name (default name is *Simple Plate.clt*), use the default type (*CL Files*), and click *Save*. Open the CL file to review its contents (Figure 2.35). The GOTO indicates how the cutter moves along the profile of the part. The X-, Y-, and Z-coordinates of the cutter location are identical to those of the G-code.

2.5   Reviewing Machining Time

From *CAMWorks* operation tree ▣ , right click *Contour Mill1* and choose *Edit Definition*.

In the *Operation Parameters* dialog box (Figure 2.36), choose the *Optimize* tab, and look for *Estimated machining time*. The total tool length of feed is 38.7in., and the feed time is 0.408 minutes.

The feed time can be manually calculated by dividing the distance that the cutter travels (see Figure 2.37) by the feedrate.

Table 2.2 Cutter travel distance in one pass with feedrate turned on

Point	Distance (in.)
AB	2.375
BC	4.75
CD	2.75
DE	2
EF	2
FG	2.75
GA	2.375
Total	19

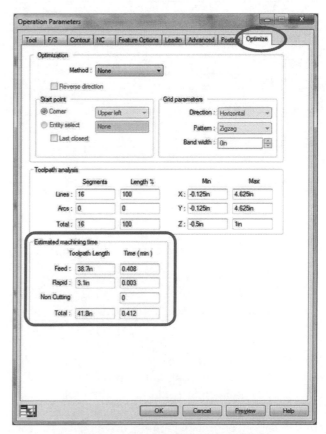

Figure 2.34 The *Optimize* tab of the *Operation Parameters* dialog box

Since the total distance *d* that the cutter travels along the part boundary profile is 38in. (19in. per pass and there are two passes), as tabulated in Table 2.2, and the feedrate *f* is 106.7 in./min. (see Figure 2.27), the feed time can be calculated as

Feed Time = d/f = 38/106.7 = 0.36 minutes

The feed time calculated is not exact since the feed in the vertical direction (for examples, N9 and N19 in Table 2.1) was not included. However, it is close to that of CAMWorks; i.e., 0.408 minutes shown in Figure 2.36.

We have completed this tutorial lesson. You may save your model for future reference.

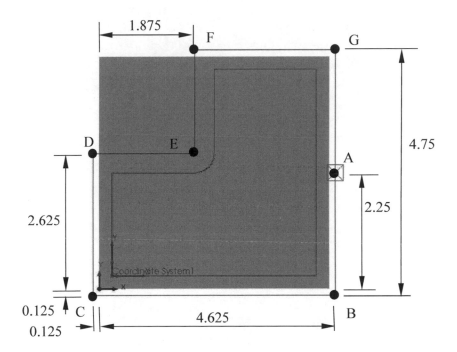

Figure 2.35 Dimensions of the cutter location points (labeled as A, B, C, and so on) and a shaded stock

2.6   Exercises

Problem 2.1. Create a contour milling operation using CAMWorks to machine a block for the design model shown in Figure 2.38. The part file can be found at the publisher's website. It is located in Lesson 2/Exercises folder. The stock is 7in.×4.5in.×0.5in. as shown in Figure 2.38. The feedrate is assumed 5 in./minute, and the depth of cut is 0.5in. In this exercise, please report the following:

- The tool you chose;
- Feed time obtained from CAMWorks;
- Feed time obtained from your own calculations;
- Screen captures for toolpath and material removal simulation in CAMWorks.

Grade should be given based on the following:

- Quality of the toolpath, i.e., the machined part must be identical to the design model (a clean cut);
- Feed time, a least possible feed time.

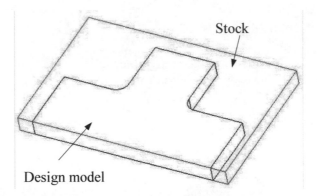

Figure 2.36 Design model and stock of Problem 2.1

Lesson 3: Machining 2.5 Axis Features

3.1   Overview of the Lesson

In this lesson, we focus on machining 2.5 axis features. One of the major characteristics of a 2.5 axis feature is that the top and bottom of the feature are flat and normal to the tool axis of the machining operations under the mill part setup. Such features include prismatic solid features and solid features with tapered walls. Typical 2.5 axis features can be a boss, pocket, open pocket, corner slot, slot, hole, face feature, open profile, curve or engrave feature. Some of these features are illustrated in Figure A.1 of Appendix A.

Features of this type are often represented as a profile sketch in CAD software that gives the height of the feature at each characteristic's point. 2.5 axis features are often preferred for machining, as it is easy to generate G-code for them in an efficient, often close to optimal fashion. In general, 2.5 axis features can be machined using a 3-axis mill.

Most 2.5 axis features, such as pockets, holes, slots, and bosses, can be extracted as machinable features by using the automative feature recognition (AFR) capability of CAMWorks. Others, such as face milling features, or contour milling features (as seen in Lesson 2), can be recognized interactively. In this lesson, we will use both methods, AFR and IFR, to extract and define respective machinable features from a solid model, generate operation plans and toolpaths, simulate and step through toolpaths, and post process the toolpaths for G-codes.

The part solid model of this lesson shown in Figure 3.1(a) consists of six holes and a pocket, which are 2.5 axis features and are extracted as machinable features automatically. In addition, we will create a face milling operation that removes a thin layer of material on top of the part. A material removal simualtion is shown in Figure 3.1(b). We will learn how to interactively create a face machinable feature from a solid model. We will follow the general steps shown in Figure 1.1 of Lesson 1 to go over this exercise.

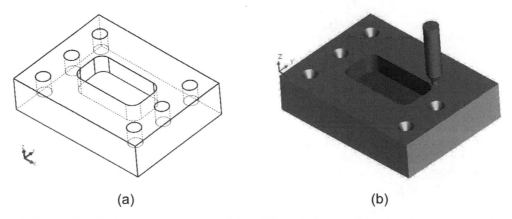

(a)   (b)

Figure 3.1 The 2.5 axis features example, (a) solid model, and (b) material removal simulation

After completing this lesson, you should be able to generate machining simulations for similar parts following the same procedures. In this exercise, we will use default options for most of the selections.

3.2   The 2.5 Axis Features Example

The size of the bounding box of the part (filename: *2 point 5 axis features.SLDPRT*) is 8in.×6in.×2in., as shown in Figure 3.2(a). The base block is created as a *Boss-Extrude1* solid feature. The size of the center pocket, *Cut-Extrude1*, is 4in.×2in.×1in. with fillets of 0.5in. in radius at the four corners; see Figure 3.2 (b). There are six blind holes, three on each side, of diameter 0.75in. and depth 1.0in.; see Figure 3.2 (c). The first hole was created as a cut extrude feature (*Cut-Extrude2*) and then duplicated for additional instances by using a pattern feature, as shown in Figure 3.2 (d). In addition, a coordinate system (*Coordinate System1*) is created at the front left corner of the bottom face of the part. Again, this coordinate system will be chosen as the fixture coordinate system, which defines the "home point" or main zero position on the machine. The unit system chosen is IPS (inch, pound, second). When you open the solid model: *2 point 5 axis features.SLDPRT*, you should see the four solid features and a coordinate system listed in the solid feature tree like that of Figure 3.3.

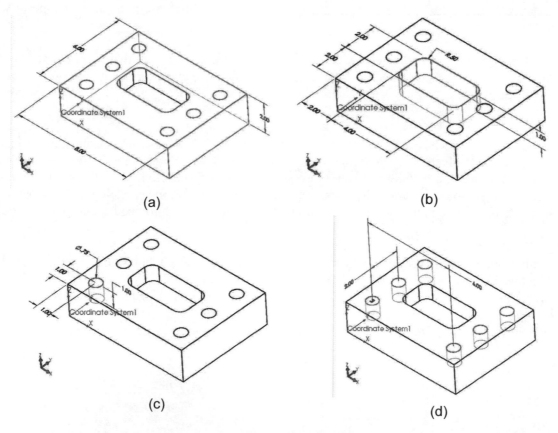

(a)   (b)

(c)   (d)

Figure 3.2 Dimensions of the solid features of the design model: (a) the base extrude, (b) the pocket, (3) the corner hole, and (d) the hole linear pattern

A stock of retangular block with size 8in.×6in.×2.25in., made of low carbon alloy steel (1005), as shown in Figure 3.4, is chosen for the machining operations. Note that a part setup origin is defined at the top left vertex of the stock, which defines the G-code program zero location. A layer of 0.25in. above the top face of the part will be removed by creating a face milling operation.

There are five NC operations to be generated for this example. The first operation is a face milling, which is a rough cut using a 1.5in. flat-end mill (T55 in the tool crib). The toolpath of the face milling is shown in Figure 3.5(a). The second operation is a rough mill operation that machines the center pocket (*Cut-Extrude1* solid feature) using a 0.5625in. flat-end mill (T54) and the third operation is a contour mill for a finish cut along the inner boundary face of the pocket using the same end mill (T54). The toolpaths of the rough mill and contour mill are shown in Figure 3.5(b). The fourth and fifth operations drill the six holes, which consist of center drill using a #8 center drill bit (T52) and hole drilling using a 3/4×118° drill (T53). The toolpath of the hole drilling operations (including the center drill) is shown in Figure 3.5(c).

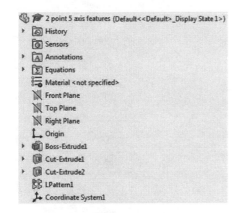

Figure 3.3 Solid feature tree of the example part

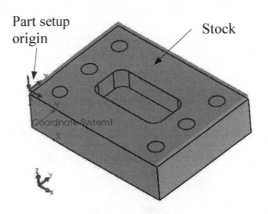

Figure 3.4 Stock with a part setup origin created at the front left vertex on top face

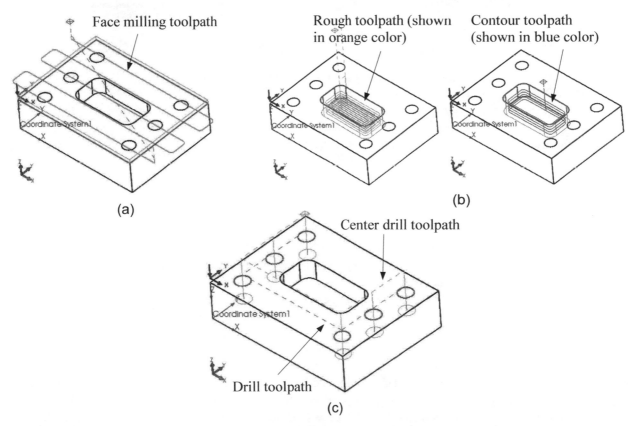

Figure 3.5 Toolpaths of the five NC operations: (a) face milling, (b) pocket milling: rough and contour, and (c) hole drilling: center drill and drill

The pocket and holes are recognized as machinable features using the automatic feature recognition (AFR), and the machinable feature for the face milling operation will be created manually. Like Lesson 2, we will use mostly the default options and parameters chosen by TechDB™.

You may open the example file with toolpaths created (filename: *2 point 5 axis feature with toolpath.SLDPRT*) to preview its toolpaths. When you open the file, you should see the five operations listed under the CAMWorks operation tree tab , as shown in Figure 3.6. You may simulate individual operations by right clicking the entity and choose *Simulate Toolpath*, or simulate the combined operations by clicking the *Simulate Toolpath* button above the graphics area.

Figure 3.6 The five NC operations listed under the CAMWorks operation tree tab

3.3   Using CAMWorks

Open SolidWorks Part

Open the part solid model (filename: *2 point 5 axis features.SLDPRT*) downloaded from the publisher's website. This solid model, as shown in Figure 3.2, consists of four solid features and a coordinate system. As soon as you open the model, you may want to check the unit system chosen and make sure the IPS system is selected. You may also increase the decimals from the default 2 to 4 digits similar to that of Lesson 2.

Enter CAMWorks

Like what we learned in Lesson 2, at the top of the feature tree you should see two important tabs: CAMWorks feature tree and CAMWorks operation tree . You should also see CAMWorks buttons on top of the graphics area (after clicking the *CAMWorks 2016* tab).

Select NC Machine

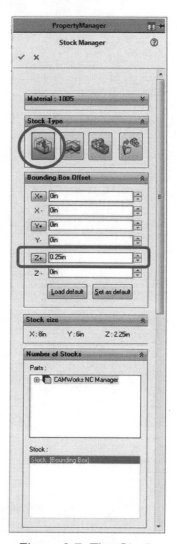

Figure 3.7 The *Stock Manager* dialog box

Click the CAMWorks feature tree tab and right click *Mill-in* to select *Edit Definition*. Similar to those of Lesson 2, in the *Machine* dialog box, we will select *Mill-in* (already selected) under *Machine* tab, choose *Crib1* under *Available tool cribs* of the *Tool Crib* tab, select *M3AXIS-TUTORIAL* under the *Post Processor* tab, and select *Coordinate System1* under *Fixture Coordinate System* of the *Setup* tab.

Create Stock

From CAMWorks feature tree ▣, right click *Stock Manager* and choose *Edit Definition*. In the *Stock Manager* dialog box (Figure 3.7), we increase the height of the stock by 0.25in. on the top side (that is, enter 0.25 for Z+, as shown in Figure 3.7). The default stock material is *Steel 1005*. Accept the revised stock by clicking the checkmark ✔ at the top left corner of the dialog box. A rectangular stock should appear in the graphics window similar to that of Figure 3.4.

Mill Part Setup and Machinable Features

Click the *Extract Machinable Features* button 📷 above the graphics area (or choose from the pull-down menu *Tools > CAMWorks > Extract Machinable Features*). A *Mill Part Setup1* is created with two machinable features extracted: *Hole Group1* (with six holes) and *Rectangular Pocket1*, all listed in CAMWorks feature tree ▣ (Figure 3.8). Both machinable features are shown in magenta color since there are no NC operations generated for these features yet.

Figure 3.8 The machinable features recognized

Figure 3.9 The operations generated

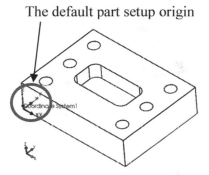

Figure 3.10 The default part setup origin coinciding with *Coordinate System1*

Generate Operation Plan and Toolpath

Click the *Generate Operation Plan* button 🔲 above the graphics area. Four NC operations: center drill, drill, rough mill, and contour mill, are generated. They are listed in CAMWorks operation tree ▣ (Figure 3.9). Again they are shown in magenta color, indicating that these opoerations are not completely defined yet. Click the *Generate Toolpath* button 📷 above the graphics area to create toolpath. The four operations are turned into a black color after toolpaths are generated.

Note that the default part setup origin coincides with the origin of the coordinate system (*Coordinate System1*), as circled in Figure 3.10, which may not be adequate for this example. As discussed in Lesson 2, the part setup origin defines the program zero location for the G-codes. We will relocate the origin to the front left corner at the top face of the stock. This corner point of the stock is easier to access at a CNC mill and is commonly chosen as the origin in carrying out machining tasks. You may select another adequate location as the G-code origin as long as the selected origin can be set up physically on the shop floor.

Relocate the origin by right clicking *Mill Part Setup1* under the CAMWorks operation tree ▣ and select *Edit Definition* (see Figure 3.11).

In the *Part Setup Parameters* dialog box (Figure 3.12), choose *Stock vertex* (under the *Origin* tab), and pick the vertex at the front left corner of the top face of the stock box, as circled in Figure 3.12. Click *OK* to accept the change. A warning box appears. Click *Yes* to the question in the warning box: *The origin or machining direction or advanced parameters has changed, toolpaths need to be recalculated. Regenerate toolpaths now?*

The toolpath will be regenerated, and the part setup origin is now moved to the top left corner of the stock, as shown in Figure 3.13.

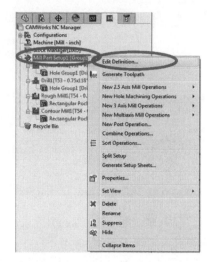

Figure 3.11

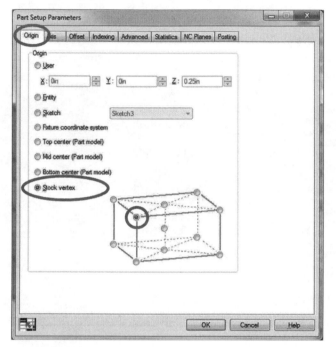

Figure 3.12 Defining the part setup origin

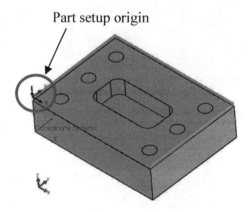

Part setup origin

Figure 3.13 The part setup origin properly relocated

You may review individual operations listed under CAMWorks operation tree ![icon] (Figure 3.9), for example, *Rough Mill1*. Right click *Rough Mill1* and choose *Edit Definition*. In the *Operation Parameters* dialog box (Figure 3.14), *T54: 9/16 EM CRB 4FL 1-1/8 LOC* is chosen. Click the *Roughing* tab; the dialog box (Figure 3.15) shows *Pocket Out* as the pocketing pattern, stepover: 50% of tool diameter, and depth parameters: 40% of the tool diameter for both the *First cut amt.* and *Max cut amt.*

You may choose other tabs, such as *F/S* (feedrate and spindle speed), to review the machining parameters determined by TechDB™. You may change these parameters and regenerate the toolpath as desired. Note that a 9/16 in. tool selected may not cut the 0.5in. fillets in the pocket. You may need to pick a tool with diameter 0.5in. or less for this operation.

Simulate Toolpath

Click the *Simulate Toolpath* button ![icon] above the graphics area. The *Toolpath Simulation* tool box appears (like that of Figure 2.28 in Lesson 2). Click the *Run* button to simulate the toolpath. The machining simulation of all four operations will appear in the graphics area, similar to that of Figure 3.16.

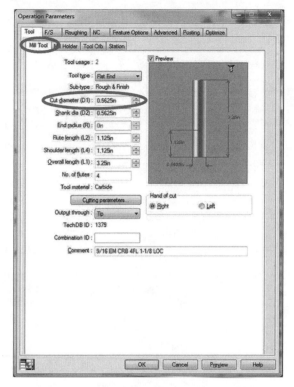

Figure 3.14 The *Mill Tool* tab of the *Operation Parameters* dialog box

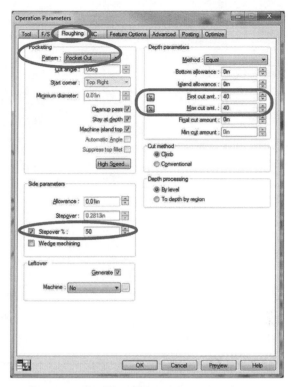

Figure 3.15 The *Roughing* tab of the *Operation Parameters* dialog box

Note that the face milling that removes the top layer material of the stock has not been created. Pocket milling and hole drilling operations may not work well in practice without the face milling completed beforehand.

Next we learn how to manually create a machinable feature for the face milling operation.

3.4 Creating a Face Milling Operation

We first tab the CAMWorks feature tree ⬛ to insert a 2.5 axis feature, and then follow the steps similar to those of Lesson 2 to create a machinable feature for face milling operation.

From CAMWorks feature tree ⬛, right click *Mill Part Setup1* and choose *New 2.5 Axis Feature* (see Figure 3.17).

Figure 3.16 Material removal simulation

In the *2.5 Axis Feature* dialog box (Figure 3.18), choose *Face Feature* for *Type*, and pick the top face of the part in the graphics area (see Figure 3.19). The top face is highlighted. The selected face (*CW Face-1*) is now listed under *Selected Entities* in the *2.5 Axis Feature* dialog box. Click the *Next* button ⬛ (circled in Figure 3.18) to define the end condition. Choose *Coarse* for *Strategy* (see Figure 3.20), and choose *Upto stock* for the *End condition: Direction 1*. Click the checkmark ✔ to accept the machinable feature.

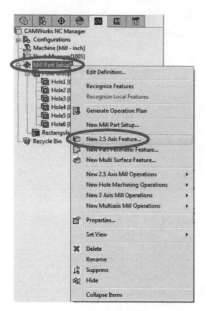

Figure 3.17 Create a new
2.5 axis feature

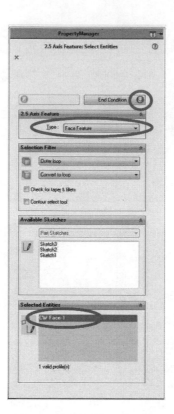

Figure 3.18 The *2.5 axis
Feature* dialog box

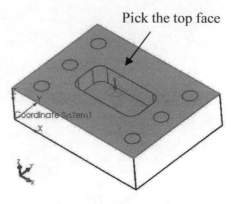

Figure 3.19 Pick the top face of the
part to create a machinable feature
for the face milling operation

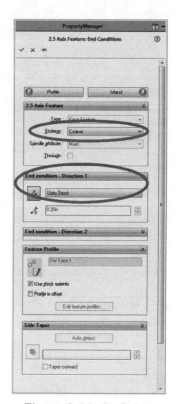

Figure 3.20 Defining
end conditions

A *Face Feature1* entity is now listed in the CAMWorks feature tree in magenta color (Figure 3.21). Right click *Face Feature1* and choose *Generate Operation Plan*. One new entity, *Face Mill1*, is now listed in CAMWorks operation tree (Figure 3.22). Right click the entity and choose *Generate Toolpath*. A face milling toolpath with a 1.5in. face mill is generated like that shown in Figure 3.23.

We will modify the face milling operation by choosing a 2in. face mill (just to learn how to add a cutter to the tool crib) and enter 0.2in. for depth parameters.

Right click *Face Mill1* in the CAMWorks operation tree and choose *Edit Definition*. In the *Operation Parameters* dialog box, choose *Tool* tab; the 1.5in. face mill is shown (Figure 3.24). Since the 2in. face mill is not in the current tool crib, we will need to add it to the crib.

First, we choose the *Tool Crib* tab and click the square box in front of the *Filter* button to display only the face mill cutters (see Figure 3.25).

Click the *Filter* button. In the *Tool Select Filter* dialog box (Figure 3.26), choose *Face Mill* for *Type*, and click the *Filter by* square box, then click *OK*. In the *Operation Parameters* dialog box, only one face mill is listed.

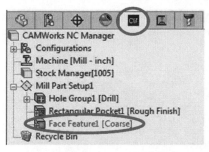

Figure 3.21 A *Face Feature1* machinable feature created

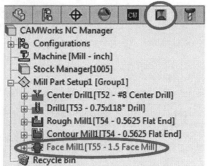

Figure 3.22 A *Face Milling1* operation generated

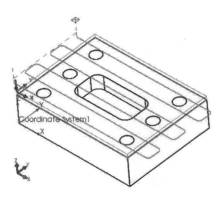

Figure 3.23 The toolpath of *Face Milling1* operation

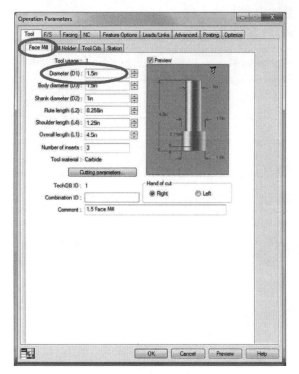

Figure 3.24 The *Face Mill* tab of the *Operation Parameters* dialog box

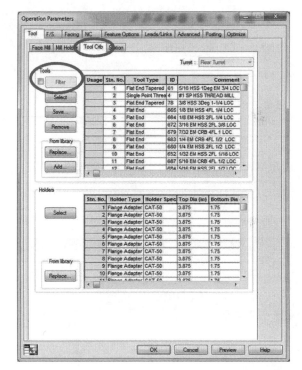

Figure 3.25 Displaying only the face mill cutters using the filter option under the *Tool Crib* tab

Click the *Add* button (see Figure 3.27) to add a face mill tool. In the *Tool Select Filter* dialog box (Figure 3.28), select *Face Mill* for *Tool type*, click the tool in the second row (ID:2) to select the 2in. face mill, and then click *OK*.

A 2in. face mill will be listed under *Tools* in the *Operation Parameters* dialog box (Figure 3.29). Choose the 2in. face mill and click *Select*. Click *Yes* to the question: *Do you want to replace the corresponding holder also?* We have now replaced the cutter with a 2in. face mill. We will modify the depth parameters next.

Choose the *Facing* tab of the *Operation Parameters* dialog box. In the *Depth parameters* group, click the percentage button ![] to deselect it for both the *First cut amt.* and *Max cut amt.* Enter *0.2in.* for both *First cut amt.* and *Max cut amt.*, as shown in Figure 3.30. Click *OK* to accept the changes. The face milling toolpath will be generated like that shown in Figure 3.5(a) with two rounds of passes, cutting 0.2in. depth in the first round and then the remaining 0.05in. in the second round.

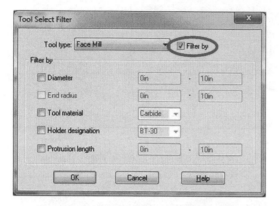

Figure 3.26 The *Tool Select Filter* dialog box

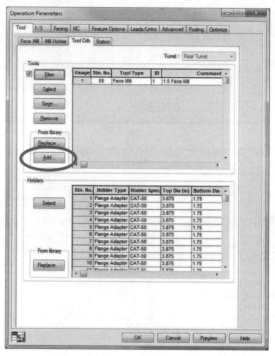

Figure 3.27 Clicking the *Add* button to add face mill under the *Tool Crib* tab

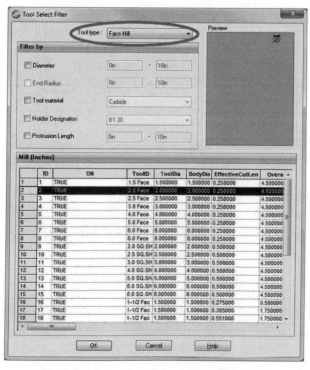

Figure 3.28 Choosing a 2in. face mill

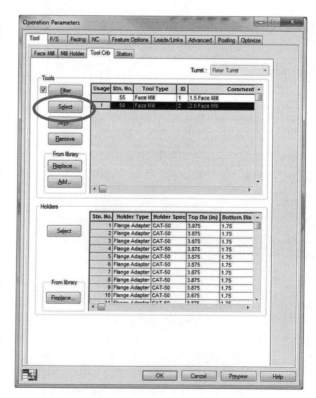

Figure 3.29 Selecting the 2in. face mill

Simulate Toolpath

Click the *Simulate Toolpath* button above the graphics area. The *Toolpath Simulation* tool box appears (see Figure 2.28 of Lesson 2). Click the *Run* button to simulate the toolpath. The material removal simulation of all five operations will appear in the graphics area at the end, similar to that of Figure 3.1.

3.5 Re-Ordering Machining Operations

Now we have created all five NC operations. However, their order is off. We would like to see these five operations in order as follows: *Face Mill1*, *Rough Mill1*, *Contour Mill1* (for the center pocket), *Center Drill1*, and *Drill1*, similar to that shown in Figure 3.31. To reorder these operations, you may click and drag them one at a time (or by pressing the shift key to select multiple) and move them up or down in the operation tree. You may click the *Simulate Toolpath* button above the graphics area to review material removal simulation for all five operations in a desired order.

3.6 Reviewing Machining Time

You may choose to review machining time for individual operations or the overall time for combined operations.

From CAMWorks operation tree, right click an operation, for example *Face Mill1*, and choose *Edit Definition*.

In the *Operation Parameters* dialog box (Figure 3.32), choose *Optimize* tab, and look for *Estimated machining time*. The estimated feed time of *Face Mill1* operation is 5.106 minutes.

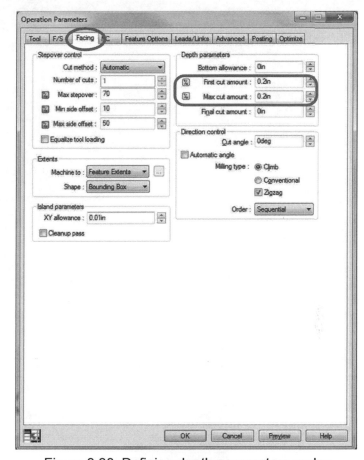

Figure 3.30 Defining depth parameters under the *Facing* tab

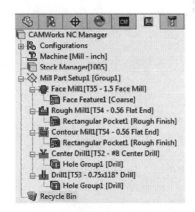

Figure 3.31 All operations in a desired order

Note that you may choose *Mill Part Setup1* and right click *Edit Definition* to review the machining time for the combined five operations. Choose *Statistics* tab in the *Part Setup Parameters* dialog box (Figure 3.33). The overall machining time for all five operations is 29.97 minutes. You may calcualte the toolpath length by sketching the toolpath (similar to that of Lesson 2) to verify the machining time.

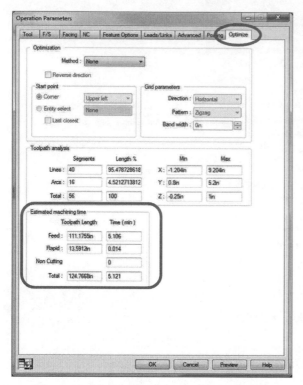

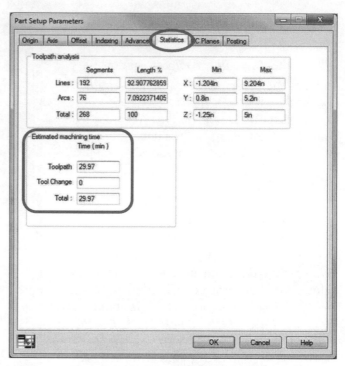

Figure 3.32 Machining time of the *Face Milling1* operation

Figure 3.33 Machining time of the combined five NC operations

3.7 The Post Process and G-Code

You may click the *Post Process* button above the graphics area, and follow the same steps learned in Lesson 2 to convert the toolpaths into G-code. Figure 3.34 shows partial contents of the G-code file. We will take a closer look at some of the NC blocks next.

Figure 3.34 Partial contents of the .txt file (G-code)

O0001	N79 Y2.8538 F10.	N249 G03 X2.2813 Y3.5 I0 J-.2188	N319 Y3.
N1 G20	N80 X5.1462	N250 G01 Y2.5	N320 Y5.
N2 (2.0 Face Mill)	N81 Y3.1463	N251 G03 X2.5 Y2.2813 I.2187 J0	N321 G80 Z1. M09
N3 G91 G28 X0 Y0 Z0	N82 X2.8538	N252 G01 X5.5	N322 G91 G28 Z0
N4 T56 M06	N83 Y3.4275	N253 G03 X5.7187 Y2.5 I0 J.2187	N323 (3/4 JOBBER DRILL)
N5 S838 M03	N84 X2.5725	N254 G01 Y3.5	N324 T53 M06
	N85 Y2.5725	N255 G03 X5.5 Y3.7188 I-.2187 J0	N325 S2000 M03
N6 (Face Mill1)	N86 X5.4275	N256 G01 X4.	
N7 G90 G54 G00 X-1.01 **Y5.2**	N87 Y3.4275	N257 G00 Z.1	N326 (Drill1)
N8 G43 Z.1 H56 M08	N88 X2.8538	N258 Z-.375	N327 G90 G54 G00 X1. Y5.
N9 G01 **Z-.2** F5.08	N89 Y3.7087	N259 G01 Z-.6688 F3.75	N328 G43 Z.1 H53 M08
N10 G17 **X0**	N90 X2.5	N260 X2.5 F30.	N329 G81 G99 R.1 Z-1.25 F10.
N11 X8. F24.1467	N91 G03 X2.2913 Y3.5 I0 J-.2088	N261 G03 X2.2813 Y3.5 I0 J-.2188	N330 Y3.
N12 X9.004	N92 G01 Y2.5	N262 G01 Y2.5	N331 Y1.
N13 G02 X9.204 Y5. I0 J-.2	...	...	N332 X7.
N14 G01 Y4.3 F150.			N333 Y3.
...	N242 (Contour Mill1)	N312 (Center Drill1)	N334 Y5.
	N243 S2200	N313 G90 G54 G00 X1. Y5.	N335 G80 Z1. M09
N74 (Rough Mill1)	N244 G41 D74 X4. Y3.7188	N314 G43 Z1. H52 M08	N336 G91 G28 Z0
N75 G90 G54 G00 X2.8538 Y3.1463	N245 Z1.	N315 G81 G98 R.1 Z-.9414 F6.	N337 G28 X0 Y0
N76 G43 Z5. H54 M08	N246 Z.1	N316 Y3.	N338 M30
N77 Z1.	N247 G01 Z-.475 F3.75	N317 Y1.	
N78 G01 Z-.475 F1.25	N248 X2.5 F30.	N318 X7.	

3.8 Stepping Through the Toolpath

We first take a closer look at the face milling toolpath by right clicking it (under the CAMWorks operation tree 🔳) and choose *Step Through Toolpath*. In the *Step Through Toolpath* dialog box (Figure 3.35), click the *Forward single step* button ▷| six times to move the cutter to X:0, Y: 5.2, and Z:–0.2 position. The cutter is in contact with the stock, as shown in Figure 3.36. This cutter location takes place after the NC block N10, in which X is 0. Z and Y positions are determined by blocks N9 and N7, respectively (see Figure 3.34). The next block (N11) moves the cutter to the front end of the stock, where X is 8, Y- and Z-coordinates stay (Y = 5.2, Z = –0.2). Note that these X-, Y-, and Z-coordinates refer to the part setup origin (again located at the front left corner of the top face of the stock), as they should be.

You may select other operations and step through the toolpaths. You will find the cutter locations (see X, Y, Z shown in the *Step Through Toolpath* dialog box, circled in Figure 3.35) are consistent with those in the G-code.

We have completed this tutorial lesson. You may want to save your model for future reference.

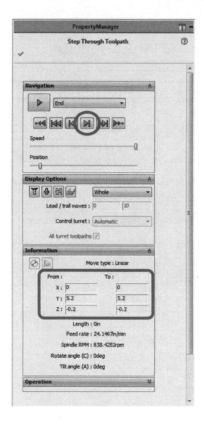

Figure 3.35 The *Step Through Toolpath* dialog box

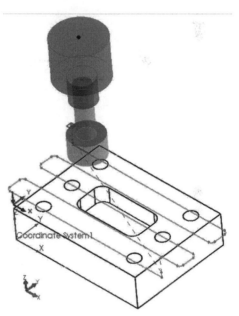

Figure 3.36 The cutter and toolpath displayed in the graphics area

3.9 Exercises

Problem 3.1. Generate machining operations to cut the part shown in Figure 3.37 from a stock of 4in.×3in.×1.25in. Pick adequate tools (with justifications). Please submit the following for grading:

(a)   A summary of the NC operations, including number of operations, cutting parameters, and tools selected.

(b)   Screen shots of combined NC toolpaths and material removal simulations.

Note that you will have to create two mill part setups to cut the features at the top and bottom of the part, respectively.

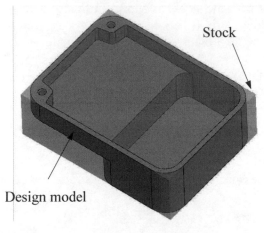

Figure 3.37 The design model and stock of Problem 3.1

Lesson 4: Machining a Freeform Surface

4.1 Overview of the Lesson

In this lesson, we focus on learning virtual machining for cutting a freeform surface (also called contoured or sculptural surface) often seen in mold or die machining. Machining a freeform surface involves rough cuts and finish cuts. Rough cuts removes a major portion of the stock material as fast as possible, often with a larger size tool and hence larger stepover and depth of cut. Multiple rough cuts may be necessary for some applications. Finish cuts polish the freeform surface to meet accuracy and quality requirements.

We will use a 3-axis mill in this lesson to cut the freeform surface. Due to a large curvature variation of the freeform surface in the example employed in this lesson, see Figure 4.1(a), a ball-nose cutter on a 3-axis mill is not able to reach certain areas for an accurate cut, which will become clear later in this lesson. We will revisit this example in Lesson 7: Multiaxis Machining by creating a multiaxis finish cut, in which a 5-axis mill is employed. Figure 4.1(b) shows the final machined part of this lesson using a 3-axis mill.

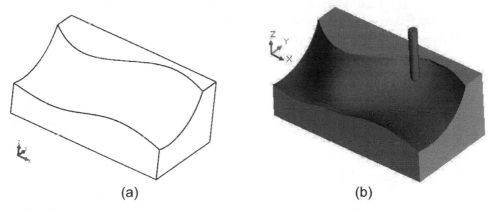

(a)   (b)

Figure 4.1 The freeform surface example, (a) solid model, and (b) material removal simulation

Since a freeform surface is not a standard 2.5 axis feature that AFR is able to extract automatically, we will learn how to manually select the surface as a machinable feature. We will then generate operation plan, generate toolpath, and simulate toolpath, similar to what we learned in previous lessons. CAMWorks generates two NC operations automatically: *Area Clearance* that removes material layer-by-layer normal to the tool axis direction, and *Pattern Project* that polishes the surface by keeping the tool tip in contact with the surface, essentially a surface milling operation.

We will take a closer look at the operations generated by CAMWorks, as well as the options and parameters determined by TechDB™. The *Area Clearance* and *Pattern Project* operations generated by CAMWorks are not quite fit for this example. The tool chosen for the *Area Clearance* operation is small; therefore, a small stepover and depth of cut are chosen, leading to an extensive machining time.

In general, machining a die with a freeform surface requires a rough cut (often called volume milling) that removes material from a raw stock as fast as possible using a larger tool, and therefore, a larger stepover and depth of cut. Since a larger tool and steps are employed, more material remains on the surface uncut at the end of the rough cut. Stock at this stage may not be ready for a finish cut that cuts the stock-in-progress closer to the finished part with desired accuracy and surface quality. Therefore, in practice we often insert a local milling in between, in which a smaller tool with smaller stepover and depth of cut are employed to remove the material remaining after the volume milling. A local milling removes the noticeable amount of material remaining on the surface and prepares the stock adequately for the final finish cut.

In this lesson we learn to implement the three-operation scenario, volume milling, local milling, and surface milling (finish cut), in CAMWorks. We first modify the *Area Clearance* operation by increasing the tool size (and stepover and depth of cut) to remove the marterial faster. We rename the operation *volume milling*. We then create a local milling (by creating another *Area Clearance*) that removes the remaining material from the first operation using a smaller cutter. Thereafter, we adjust a few parameters and try to improve the surface finish of the surface milling operation (*Pattern Project*) since the default settings do not provide a desired cut. In Lesson 7, we revisit this example by creating a multiaxis surface milling operation for a more desireable surface finish.

4.2   The Freeform Surface Example

The size of the bounding box of the part (filename: *Freeform Surface.SLDPRT*) is 7.5in.×4in.×3in. The solid model is lofted from four parallel sketches, each spaced 2.5in. apart, as shown in Figure 4.2(a). The freeform surface is formed by lofting the circular arcs of respective sketches. Individual sketches are created by four straight lines and a circular arc. The front and rear end faces of the loft share the same sketch (that is, Sketch1 and Sketch4 shown in Figure 4.2(a) are identical). Dimensions of individual sketches are shown in Figure 4.2(b), (c), and (d), respectively.

In addition, a fixture coordinate system (*Coordinate System1*) is defined at the front left corner of the bottom face of the part. Again this coordinate defines the "home point" or main zero position on the machine. The unit system chosen is IPS (inch, pound, second). When you open the solid model *Freeform Surface.SLDPRT*, you should see the loft solid feature and a coordinate system listed in the feature tree like that of Figure 4.3.

A stock of retangular block with a size 7.5in.×4in.×3in., made of low carbon alloy steel (1005), as shown in Figure 4.4, is chosen for the machining operations. Note that a part setup origin is defined at the front left vertex on the top face of the stock, which defines the G-code program zero location.

As mentioned above, we implement the three-operation scenario for this example. The first operation is a volume milling (*Area Clearance*), which is a rough cut using a 1in. hog nose mill with 0.25in. corner radius (T52 in tool crib). The toolpath of the volume milling operation is shown in Figure 4.5(a). The second operation is a local milling (another *Area Clearance*) that continues removing the material remaining from volume milling using a smaller ball nose cutter of diameter 0.5in. (T54) (see toolpath in Figure 4.5(b)). The third operation is a surface milling (*Pattern Project*) using a 0.375in. tool (T53), serving as a finish cut that is supposed to produce a machined part that meets the accuracy and quality requirements; see toolpath in Figure 4.5(c).

You may open the example file with toolpath created (filename: *Freeform Surface with toolpath.SLDPRT*) to preview the toolpath in this example. When you open the file, you should see the three operations listed under the CAMWorks operation tree tab 🖳 plus a *Multiaxis Surface Milling* suppressed (we will discuss this multiaxis surface milling operation in Lesson 7), as shown in Figure 4.6. You may simulate individual operations by right clicking an entity and choosing *Simulate Toolpath*, or simulate the combined operations by clicking the *Simulate Toolpath* button 🖥 above the graphics area.

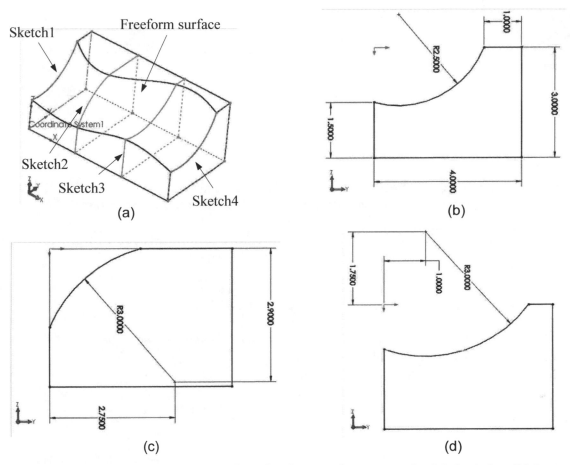

Figure 4.2 Sketches and dimensions of the freeform surface example: (a) the loft solid feature, (b) Sketch 1 (and 4), (3) Sketch 2, and (d) Sketch 3

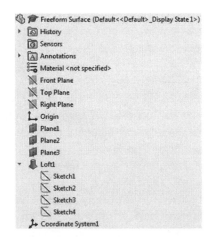

Figure 4.3 Solid feature tree of the freeform surface example

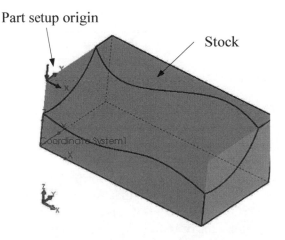

Figure 4.4 Stock with a part setup origin created at the front left vertex on top face

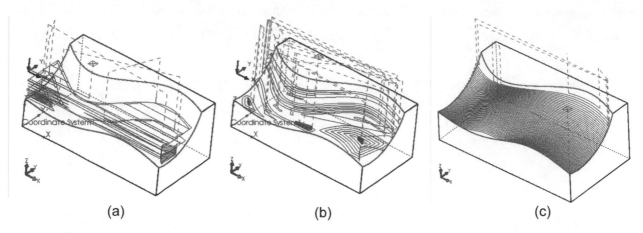

(a)   (b)   (c)

Figure 4.5 Toolpaths of the three NC operations: (a) volume milling (*Area Clearance*), (b) local milling (another *Area Clearance*), and (c) surface milling (*Pattern Project*)

4.3   Using CAMWorks

Open SOLIDWORKS Part

Open the part file (filename: *Freeform Surface.SLDPRT*) downloaded from the publisher's website. This solid model, as shown in Figure 4.2(a), consists of one loft solid feature and a coordinate system. As soon as you open the model, you may want to check the unit system chosen and make sure the IPS system is selected. You may also increase the decimals from the default 2 to 4 digits similar to that of previous lessons.

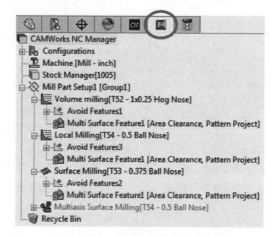

Figure 4.6 The NC operations listed under the CAMWorks operation tree tab

Select NC Machine

Click the CAMWorks feature tree tab [CW] and right click *Mill-in* to select *Edit Definition*. Similar to those of Lesson 3, in the *Machine* dialog box, we select *Mill-in* (already selected) under *Machine* tab, choose *Crib1* under *Available tool cribs* of the *Tool Crib* tab, select *M3AXIS-TUTORIAL* under the *Post Processor* tab, and select *Coordinate System1* under *Fixture Coordinate System* of the *Setup* tab.

Create Stock

From CAMWorks feature tree tab [CW], right click *Stock Manager* and choose *Edit Definition*. In the *Stock Manager* dialog box, we choose the default stock size (7.5in.×4in.×3in.) and material (*Steel 1005*). The rectangular stock should appear in the graphics window similar to that of Figure 4.4.

We will first select the freeform surface to define a machinable feature. We will let CAMWorks technology database determine machining operations, and then modify them in hope to generate a better machined surface with reduced machining time. We will relocate the part setup origin to the front left corner at the top face of the stock, as shown in Figure 4.4.

Create a Machinable Feature

Click the CAMWorks feature tree tab ██ , right click *Stock Manager*, and choose *New Mill Part Setup*, as shown in Figure 4.7.

The *Mill Setup* dialog box appears and the *Front Plane*, perpendicular to the Z-axis, is selected (if not, select *Front Plane*). We choose *Front Plane* to define tool axis of the setup. The *Front Plane* should appear under *Entity* in the dialog box. In the graphics area (see Figure 4.8), a symbol ⌖ with arrow pointing upward appears. This symbol indicates the tool axis (or feed direction) that must be reversed (pointing downward). Click the *Reverse Selected Entity* button ↗ under *Entity* to reverse the direction. Make sure that the arrow points in a downward direction. Click the checkmark ✓ to accept the definition. A *Mill Part Setup1* is now listed in the feature tree.

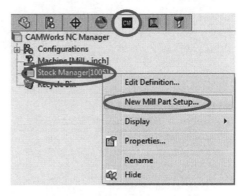

Figure 4.7 Selecting *New Mill Part Setup*

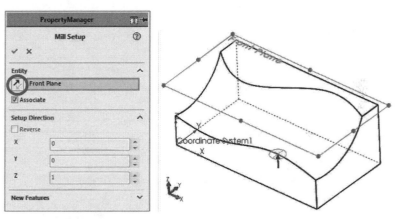

Figure 4.8 Pick the top face for mill setup

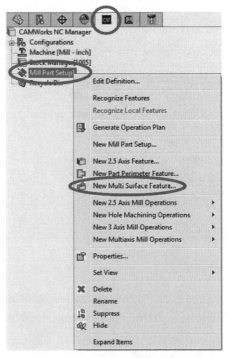

Figure 4.9 Choosing *New Multi Surface Feature*

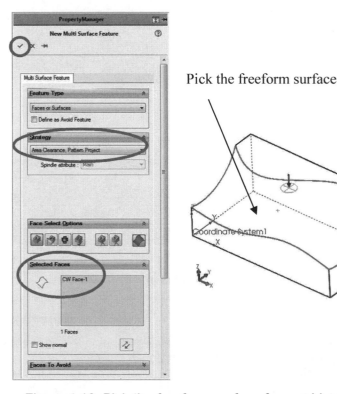

Pick the freeform surface

Figure 4.10 Pick the freeform surface for machinable feature

Now we define a machinable feature. From CAMWorks feature tree 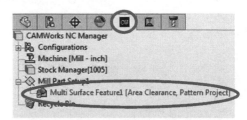, right click *Mill Part Setup1* and choose *New Multi Surface Feature* (see Figure 4.9).

In the *New Multi Surface Feature* dialog box (Figure 4.10), pick the freeform surface of the part in the graphics area (Figure 4.10); the surface picked is now listed under *Selected Faces*. Leave the default *Area Clearance, Pattern Project* for *Strategy*, and click the checkmark to accept the surface feature.

In the CAMWorks feature tree 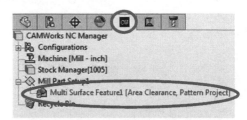, a *Multi Surface Feature1* is added in magenta color under *Mill Part Setup1*, as shown in Figure 4.11.

Next we create another multi surface feature and pick the top face of the part to define it as the area to avoid. This is to restrict the toolpath to stay only on the freeform surface.

Right click *Mill Part Setup1* and choose *New Multi Surface Feature*.

Figure 4.11

Pick the top face to avoid

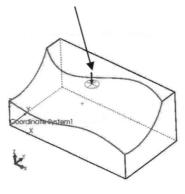

Figure 4.13 The two operations generated

Figure 4.12 Pick the top face as a face to avoid

In the *New Multi Surface Feature* dialog box, pick the top face of the part in the graphics area (see Figure 4.12), click the *Define as Avoided Feature* box, and click the checkmark to accept it.

In the CAMWorks feature tree 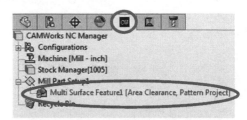, a *Multi Surface Feature2[Avoid]* is added in magenta color under *Mill Part Setup1*.

Generate Operation Plan and Toolpath

Click the *Generate Operation Plan* button ☐ above the graphics area. Two operations, *Area Clearance1* and *Pattern Project1*, are generated. They are listed in CAMWorks operation tree ☐ (as seen in Figure 4.13). Again they are shown in magenta color, indicating that these operations are not completely defined yet. If you expand an operation (for example, *Area Clearance1*) in the feature tree, an

Avoid Feature1 entity is listed. Expand the *Avoid Feature*; the *Avoid Group 1 (0)* appears, indicating that the feature to avoid has not been selected. We will accept the operations as they are for now. We will come back shortly to revisit the avoid feature selection.

Note that when you click the *Mill Part Setup* (under CAMWorks operation tree 🄲), the default part setup origin appears at the center of the top face of the stock, as circled in Figure 4.14, which is less desirable. We will relocate the origin to the front left corner at the top face of the stock.

Redefine the origin by right clicking *Mill Part Setup1* in the CAMWorks operation tree 🄲 . In the *Part Setup Parameters* dialog box (see Figure 4.15), choose *Stock vertex* (under the *Origin* tab), and pick the vertex at the front left corner of the top face. Click *OK* to accept the change. The part setup origin is now moved to the front left corner of the top face of the stock like that of Figure 4.4.

The default part setup origin

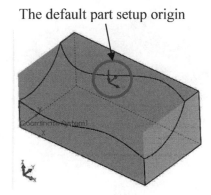

Figure 4.14 The default part setup origin located at the center of the stock top face

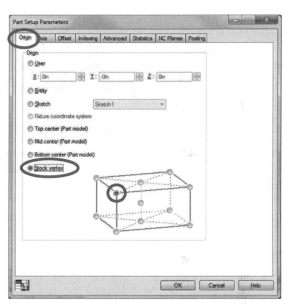

Figure 4.15 Selecting vertex in the *Part Setup Parameters* dialog box

Undesirable toolpath (behind the freeform surface)

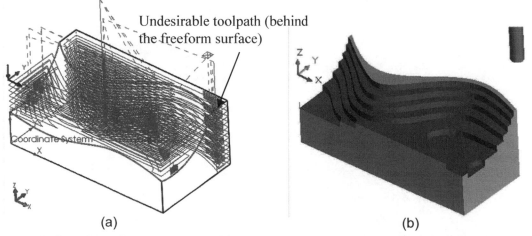

(a)   (b)

Figure 4.16 The *Area Clearance1* operation, (a) toolpath, and (b) material removal simulation

Click the *Generate Toolpath* button 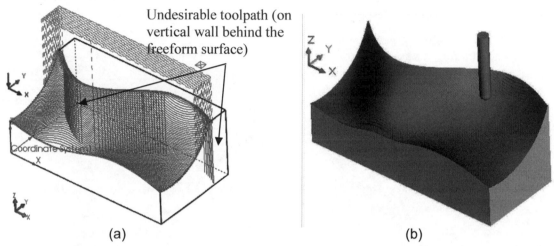 above the graphics area to create the toolpath. The two operations are turned into black color right after toolpaths are generated.

Toolpaths of the two operations, *Area Clearance* and *Pattern Project*, are generated like those in Figure 4.16(a) and Figure 4.17(a), respectively. The toolpaths appear not only removing material above the freeform surface, but cutting the area below the top face of the part (behind the freeform surface). The material removal simulations—shown in Figure 4.16(b) and Figure 4.17(b)—confirm the case. These toolpaths are certainly not desirable. We need to contain the toolpaths to only machine the freeform surface by selecting the top face of the part as the avoid feature.

(a)   (b)

Figure 4.17 The *Pattern Project1* operation, (a) toolpath, and (b) material removal simulation

4.4   Creating Avoid Feature to Correct the Toolpath

Right click *Avoid Group 1(0)* under *Avoid Feature1* of *Area Clearance1* and select *Edit Definition* (see Figure 4.18).

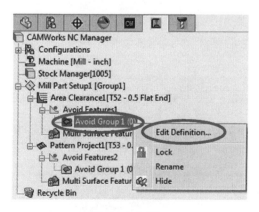

Figure 4.18

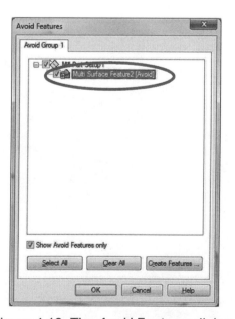

Figure 4.19 The *Avoid Features* dialog box

Choose *Multi Surface Feature2[Avoid]* in the *Avoid Features* dialog box (Figure 4.19), then click *OK*.

Repeat the same steps for the *Pattern Project1* operation.

Click the *Generate Toolpath* button above the graphics area. The toolpaths will be regenerated like that shown in Figure 4.20. Note that the toolpaths behind the freeform surface disappear as desired.

Simulate the toolpaths (click the *Simulate Toolpath* button above the graphics area). The material removal simulation shown in Figure 4.21(a) seems to indicate that the machining operations cut a desirable freeform surface from a rectangular stock. Is this the case? Let us take a closer look.

In the *Toolpath Simulation* tool box, Figure 4.21(b), select the *Show difference* button—circled in Figure 4.21(b)—to show a visual comparison of the machined part and the design part in the graphics area; see Figure 4.21(c). As shown in Figure 4.21(c), the cut is not clean; for example, the area close to the middle of the front edge circled in Figure 4.21(c). Other areas, such as those close to the top edge, show similar issues.

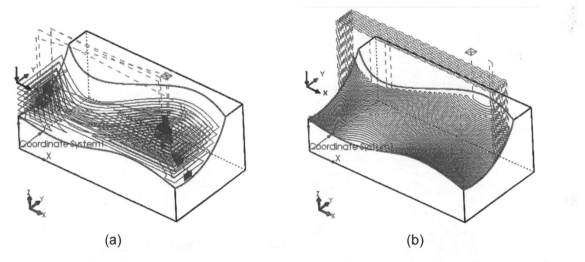

(a)   (b)

Figure 4.20 Toolpaths with avoid features selected, (a) *Area Clearance1*, and (b) *Pattern Project1*

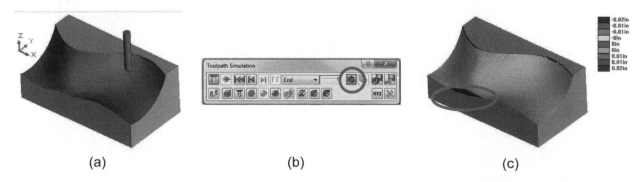

(a)   (b)   (c)

Figure 4.21 The material removal simulation, (a) combined operations, (b) the *Show Difference* button of the *Toolpath Simulation* tool box, and (c) differences between the machined part and the design part shown with a color spectrum

How long does the entire machining operation take? The machining times are 111.1 and 17.3 minutes, respectively, for the two operations, based on the feedrates determined by the TechDB™. Can the machining time of the *Area Clearance1* be reduced? Is the toolpath of *Pattern Project* fine enough to produce a satisfactory finished surface? Can we modify the toolpath of the rough cut (*Area Clearance*) with reduced machining time and improve the toolpath for a better surface finish (*Pattern Project*) with better accuracy [remember the areas that show significant deviation between the finished part and the design part in Figure 4.21(c)]?

4.5   Modifying the Area Clearance Toolpath

How do we modify the toolpath? Where to find and modify key parameters that revise the toolpath, such as depth of cut, stepover, and cut pattern? Let us first take a closer look at the first toolpath, *Area Clearance1*. We would like to use a larger size tool and increase stepover and depth of cut to reduce the machining time.

In CAMWorks operation tree 📐, right click *Area Clearance1* and choose *Edit Definition*. In the *Operation Parameters* dialog box, Figure 4.22(a), T52 has been selected (0.5in. flat-end mill) under the *Tool* tab. We will use a 1in. hog nose cutter with a corner radius 0.25in. for this operation.

In the *Operation Parameters* dialog box, choose *Tool Crib* tab, and only display hog nose cutters using the filter option (see Lesson 3 for details).

Choose the 1in. hog nose cutter with corner radis 0.06in.—*ID:631* in Figure 4.21(b). Change the corner radius to 0.25in., and click *Select*. Click *Yes* to the question: *Do you want to replace the corresponding holder also?* We have now repalced the tool with a modified 1in. hog nose cutter. We will modify the machining parameters next.

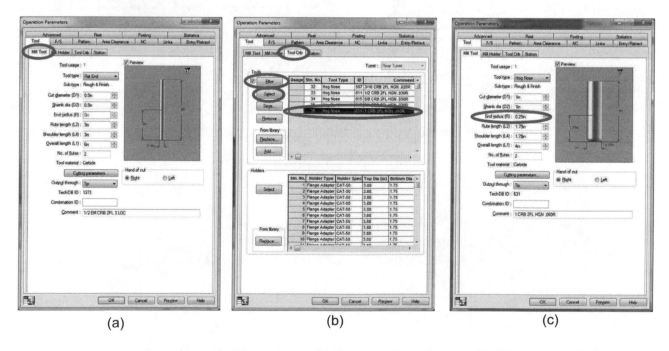

(a)   (b)   (c)

Figure 4.22 Replacing the tool with a 1in. hog nose, (a) 0.5in. flat-end mill selected currently, (b) showing hog nose cutters in the tool crib using the filter option, and (c) 1in. hog nose cutter selected

Choose the *Pattern* tab of the *Operation Parameters* dialog box, select *Lace* for *Pattern*, and enter 90% for *Lace stepover*, as shown in Figure 4.23(a).

Choose the *Area Clearance* tab of the *Operation Parameters* dialog box; see Figure 4.23(b). In the *Depth* parameters group, enter *0.5in.* for *Cut amount*, as shown in Figure 4.23(b). Click *OK* to accept the changes and click *Yes* to the warning message: *Operation parameters have changed, toolpaths need to be recalculated. Regenerate toolpaths now?*

The area clearance toolpath will be generated like that shown in Figure 4.5(a). The machining time is now 67.9 minutes (click the *Statistics* tab to see the time), as shown in Figure 4.23(c). Although the machining time is reduced, is the operation acceptable and is the stock-in-progress ready for the next operation, *Pattern Project*?

Right click *Area Clearance1* and choose *Rename*. Change the name of the operation to *Volume Milling*.

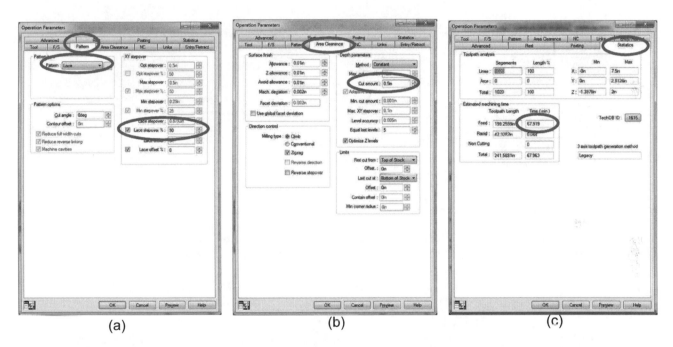

(a)   (b)   (c)

Figure 4.23 Modifying machining parameters, (a) edit parameter/option under the *Pattern* tab, (b) change *Cut amount* under the *Area Clearance* tab, and (c) show machining time in the *Statistics* tab

4.6   Reviewing the Machined Part Quality

Now let us take a closer look at the machined part after carrying out the *Volume Milling* operation. We will learn to use a few more buttons in the *Simulate Toolpath* tool box, in particular the *Section view* button to create section views.

Right click *Volume Milling* and select *Simulate Toolpath*. The *Toolpath Simulation* tool box appears; see Figure 4.24(a). Click the *Run* button ▶ to simulate the toolpath. The material removal simulation of the *Area Clearance1* operation will appear in the graphics area at the end, similar to that of Figure 4.24(a).

Click the *Stock Display* button  and choose *Wireframe Display* to display the stock in wireframe; see Figure 4.24(b), which shows a better view in terms of the geometric shape of the machined surface in progress.

You may click the *Tool Display* button and choose *No Display* to turn off the tool display, and click the *Target Part Display* button and choose *Shaded Display* to show a shaded display of the design (or target) part overlapping with the machined part or a stock-in-progress; see Figure 4.24(c). Boundary edges of the design part shown in yellow color clearly differentiate the significant amount of material that remained after the *Volume Milling* operation.

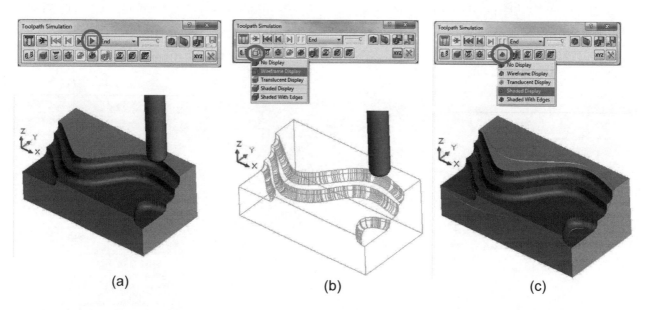

(a)   (b)   (c)

Figure 4.24 Options of the *Simulate Toolpath* tool box, (a) material removal simulation, (b) stock displayed in wireframe, and (c) tool display turned off and target part in shaded display

Figure 4.25 The *Toolpath Simulation* tool box

Figure 4.26 The *Section view* dialog box

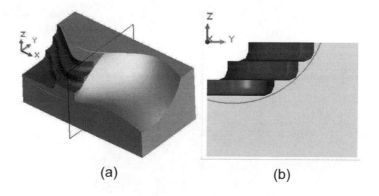

(a)   (b)

Figure 4.27 Section views of YZ plane, (a) current work view, and (b) right end view

You may create a section view to review the quality of the machined part at specific sections normal to a direction of a selected plane. Click the *Section view* button ![] from the *Toolpath Simulation* tool box (Figure 4.25). In the *Section view* dialog box, choose *YZ* for *Plane*, click the *Reverse direction* button ![], and increase the *Offset* to *3.3in.* (Figure 4.26). A section view appears like that of Figure 4.27(a), which shows the difference between the target and the machined parts at the specific section. You may rotate the view to see the section, for example, viewing it from the right end; see Figure 4.27(b). It is apparent that there is a significant amount of material uncut. Next, we create a local milling operation to remove the uncut material with a smaller tool and reduced stepover and depth of cut.

4.7   Adding a Local Milling Operation

We will add a local milling operation by creating another area clearance operation to continue cutting the material remaining from the *Volume Milling* operation with a ball cutter of 0.5in. diameter. We enter 0.2in for both stepover and depth of cut.

Under the CAMWorks operation tree ![], right click *Mill Part Setup1* and choose *New 3 Axis Mill Operations > Area Clearance*.

The *New Operation: Area Clearance* dialog box (Figure 4.28) appears. Click the CAMWorks feature tree tab ![] and select *Multi Surface Feature1* (Figure 4.29). The selected entity is listed under *Features* of the *New Operation: Area Clearance* dialog box (see Figure 4.28). Click the checkmark ✔ to accept the new operation.

In the *Operation Parameters* dialog box, choose *Tool Crib* tab, and only display ball nose cutters using filter option. Choose the 0.5in. cutter (*ID: 786, 1/2 HSS 2FL*) as seen in Figure 4.30(a), and click *Select*. Click *Yes* to replace the tool holder.

Choose the *Pattern* tab of the *Operation Parameters* dialog box, select *Pocket Out* for *Pattern*, enter 40% and 20% for *Max stepover%* and *Min stepover%*, as shown in Figure 4.30(b).

Choose the *Area Clearance* tab of the *Operation Parameters* dialog box; see Figure 4.30(c). In the *Depth* parameters group, enter *0.2in* for *Cut amount*, as shown in Figure 4.30(c).

Choose the *Rest* tab of the *Operation Parameters* dialog box, as shown in Figure 4.31(a). Choose *From WIP* for *Method*, and click the selection button ![] to select *Volume Milling* for *Compute WIP from operations*; see Figure 4.31(a). This is how we link the local milling to the previous operation(s). WIP stands for work in progress. Click *OK* to accept the changes.

Expand the newly created operation (*Area Clearance2*) to verify that the number in the parentheses in *Avoid Group 1* is not zero. If it is, right click *Avoid Group 1* to add *Multi Surface Feature2* as the aviod feature (see Figure 4.19).

Figure 4.28 The *New Operation: Area Clearance* dialog box

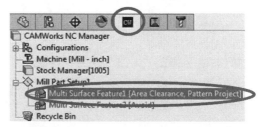

Figure 4.29 Select *Multi Surface Feature1*

Right click *Area Clearance2* and choose *Generate Toolpath* to generate a toolpath like that shown in Figure 4.5(b).

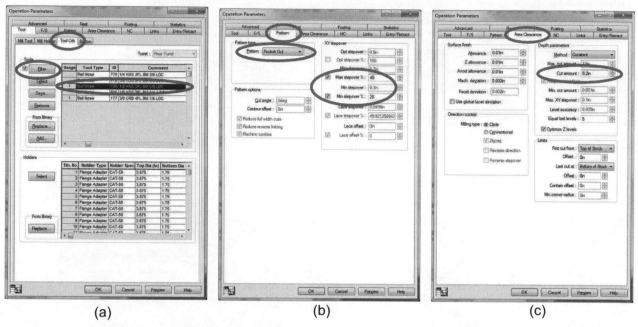

(a)   (b)   (c)

Figure 4.30 Modifying cutter and machining parameters, (a) filter to display ball nose cutters under the *Tool Crib* tab, (b) edit parameter/option under the *Pattern* tab, and (c) change cut amount under the *Area Clearance* tab

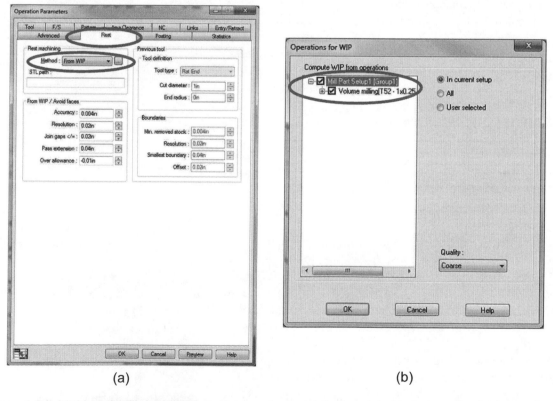

(a)   (b)

Figure 4.31 Linking the local milling to the previous operation, (a) choose method under the *Rest* tab, and (b) pick *Volume Milling* in the *Operations for WIP* dialog box

Right click again *Area Clearance2* and choose *Rename*. Change the name of the operation to *Local Milling*. The machining time for the *Local Milling* operation is 39.3 min., calculated using the default feedrate chosen by TechDB™.

Simulate the toolpaths for *Volume Milling* and *Local Milling* by, for example, supressing *Pattern Project* (right click *Pattern Project* and choose *Supress*) and clicking the *Simulate Toolpath* button [Simulate Toolpath] above the graphics area. Click the *Run* button [▶] to simulate the toolpath. The material removal simulation of the two operations appear in the graphics area, similar to that of Figure 4.32.

By visually comparing the material removal simualtions of the combined *Volume Milling* and *Local Milling* with that of *Volume Milling* only (shown in Figure 4.33), the quality of the machined part by the combined operations has less material remaining, as can be seen at the section views.

The machining time of the combined operations is 107.2 min. (39.3 + 67.9), slightly less than that of *Volume Milling* (111.1 min.), all calculated using default feedrates chosen by TechDB™.

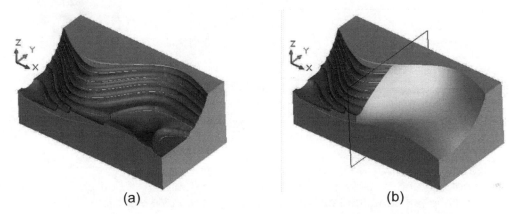

(a)   (b)

Figure 4.32 Material removal simulation for *Volume Milling* and *Local Milling*, (a) current work view, and (b) section view

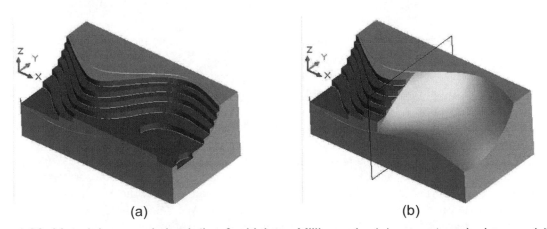

(a)   (b)

Figure 4.33 Material removal simulation for *Volume Milling* only, (a) current work view, and (b) section view

4.8   Reviewing and Modifying the Pattern Project Operation

Under CAMWorks operation tree ![icon], click *Pattern Project* to show its toolpath; see Figure 4.34(a). Note that the toolpath was generated based on the *Slice* pattern type and a stepover 0.05in. The toolpath seems to be fine except that the toolpath near the front edge—circled in Figure 4.34(a)—may leave a noticeable amount of material uncut. A similar observation is noted on the top edge of the freeform surface.

Now, we simulate the toolpaths for all three operations combined: *Volume Milling*, *Local Milling*, and *Pattern Project*.

Click the *Simulate Toolpath* button ![button] above the graphics area. Click the *Run* button ![run] to simulate the operations. The material removal simulation of the combined operations appears in the graphics area, similar to that of Figure 4.34(b) and Figure 4.34(c) (section view). A noticeable amount of material uncut at the front edge is apparent, as predicted. Same is true near the top edge, but less severe.

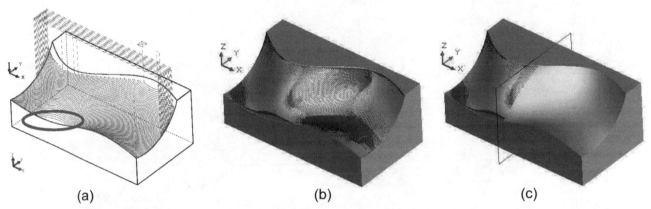

| (a) | (b) | (c) |

Figure 4.34 Toolpath and material removal simulation, (a) toolpath of *Pattern Project*, (b) material removal simulation of combined operations, and (c) section view at offset 3.2in. in YZ plane

How do we adjust the toolpath of the *Pattern Project* operation to improve the accuracy of the machined part? We will explore more on altering the options and parameters of the operation to see if an improved operation is possible.

We will explore a different pattern type; more specifically, the flowline pattern, and alter the direction of the toolpath to see if the uncut areas can be minimized.

Under CAMWorks operation tree ![icon], right click *Pattern Project* and choose *Edit Definition*. In the *Operation Parameters* dialog box, click the *Pattern* tab, as shown in Figure 4.35(a), to review the pattern type and stepover. Change the pattern to *Flowline* and change the step over to 0.05 in., which is small.

Click *Curve 1*, the *Curve Wizard*, as shown in Figure 4.35(b), appears. Select *Single Edge*, pick the top edge curve of the freeform surface (Figure 4.35(c)) in the graphics area. A dot appears on the front end of the edge curve; see Figure 4.35(c). This dot is considered the end point of the edge curve since we expect the toolpath to start on the rear edge of the freeform surface. We select the *End of Curve* under *Start point*. As soon as the *End of Curve* under *Start point* is selected, the point at the rear end of the curve (*Curve 1*) is highlighted, indicating the point is the start point of the edge curve.

Click *Curve 2*, and pick the bottom edge curve of the freeform surface in the graphics area. Notice that the point at the rear end of the curve (*Curve 2*) is highlighted, indicating the point is the start point of the curve. This ensures that the start points of both curves are consistent leading to a valid toolpath.

Leave *Along* for *Cut*, and click the *Preview* button to show the toolpath like that of Figure 4.36(a). Note that the toolpath still leaves a noticeable gap at the front edge (and rear edge) as circled in Figure 4.36(a). We accept the toolpath for the time being. Click the close button ⊠ on the top right corner of the *Operation Parameters* dialog box—circled in Figure 4.35(a)—and click *OK* to accept the changes and regenerate the toolpath.

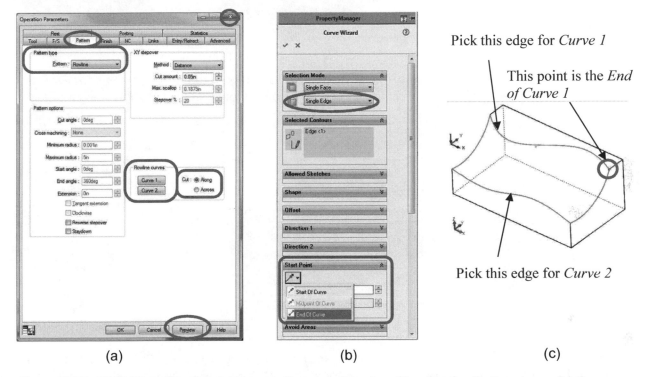

(a)   (b)   (c)

Figure 4.35 Defining a *Flowline* type operation, (a) choosing *Flowline* for *Pattern* type, (b) the curve wizard showing single edge, curve picked, and options of selecting start point, and (c) picking curves

Simulate the toolpaths of three combined operations, as shown in Figure 4.36(b), and use section view to review the machined surface; see Figure 4.36(c).

The material removal simulation indicates that there is still a noticeable amount of material remaining on the freeform surface, especially near the front edge in the area circled in Figure 4.36(c).

Now we alter the toolpath direction by choosing *Across* for *Cut* in the *Pattern* tab of the *Operation Parameters* dialog box; see Figure 4.35(a). The toolpath still leaves a noticeable gap at the front edge (and rear edge) as circled in Figure 4.37(a). The material removal simulation indicates the same; see Figure 4.37(b) and (c).

With a 3/8in. tool and 3-axis mill, the curvature of the convex area makes the surface too steep for the tool to reach, leading to the noticeable uncut material. This problem can be properly addressed using a 5-axis mill. We will learn to generate operations using 5-axis mill in Lesson 7, and revisit this example at the time.

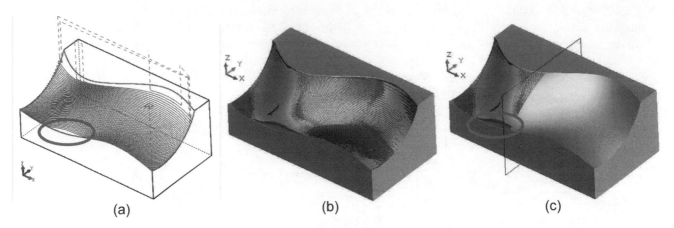

Figure 4.36 Toolpath and material removal simulation, (a) toolpath of the revised *Pattern Project* operation, (b) material removal simulation of combined operations, and (c) section view at offset 3.2in. in YZ plane

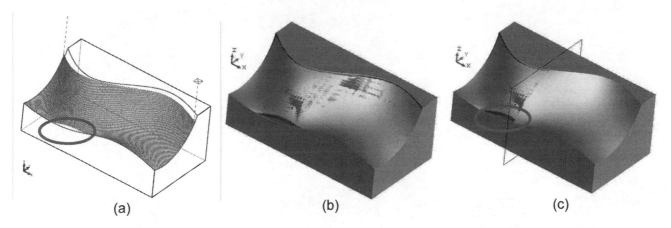

Figure 4.37 Toolpath and material removal simulation (choosing *Across* for *Cut*), (a) toolpath, (b) material removal simulation of combined operations, and (c) section view at offset 3.2in. in YZ plane

We have completed this exerciese. You may save your model for future reference. Please keep the model file since we will need it in Lesson 7.

4.9   Exercises

Problem 4.1. Follow the same steps discussed in this lesson to machine a part shown in Figure 4.38 using a stock of 12in.×5.5in.×3.5in. Note that at least three operations using a 3-axis mill must be included: volume milling, local milling, and surface milling. Report the tools and machining options selected for individual operations. Also report machining times of individual and combined operations. Would operations of 3-axis mill give you a satisfactory machined part? Do you notice uncut areas like those in, for example, Figure 4.36?

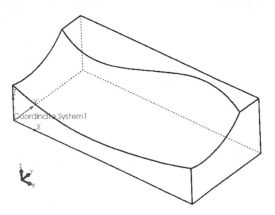

Figure 4.38 Part of Problem 4.1

Lesson 5: Multipart Machining

5.1   Overview of the Lesson

So far, we have discussed virtual machining for cutting a single part that was created as a solid part in SOLIDWORKS. In this lesson, we will move one step further, in which we focus on creating machining simulations that machine a set of identical parts in an assembly created in SOLIDWORKS. Machining multiple parts in a single setup is a common practice at shop floor in industry.

The individual part to be machined is identical to that in Lesson 3, which involves pocket milling and hole drilling using a 3-axis mill. The face milling is not included in this lesson to simplify fixture design that holds the stock to a jig table. In SOLIDWORKS assembly, the part (more precisely, the stock) is assembled to a jig table by using two fixtures (one on each side). Each fixture consists of a clamp, a riser, and a bolt. A total of ten parts in two rows are to be machined, as shown in Figure 5.1.

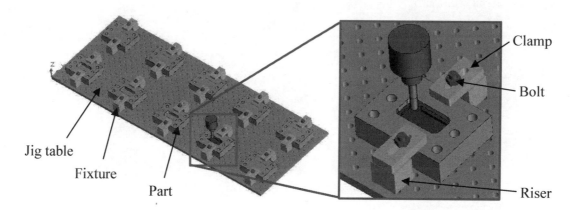

Figure 5.1 The material removal simulation of the multipart machining example

In this lesson, we will learn the steps to create instances of the part in CAMWorks, define stocks for individual instances, extract machinable features, generate toolpath, and select components in the assembly (for example, the jig table) for the tools to avoid. In addition, we will take a closer look at the G-code generated by CAMWorks for machining multiple parts in an assembly.

5.2   The Multipart Machining Example

The size of the bounding box of the part (filename: *2 point 5 axis features.SLDPRT*) is 8in.×6in.×2in., as shown in Figure 5.2(a). There is a center pocket and six holes, three on each side. As discussed in Lesson 3, these are 2.5 axis features that can be extracted as machinable features by using the automatic feature recognition (AFR) capability. The stock size is identical to that of the bounding box. As mentioned earlier, the face milling operation discussed in Lesson 3 is excluded in this lesson. The pocket milling operations, including a rough mill and a contour mill—toolpaths shown in Figure 5.2(b)—and hole drilling operations—center drill and drill shown in Figure 5.2(c)—are identical to those of Lesson 3.

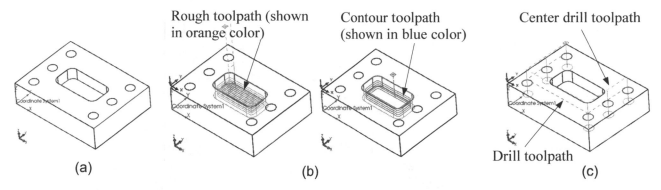

Figure 5.2 The *2 and 5 axis features* example: (a) solid model, (b) pocket milling toolpaths: rough and contour, and (c) hole drilling toolpaths: center drill and drill

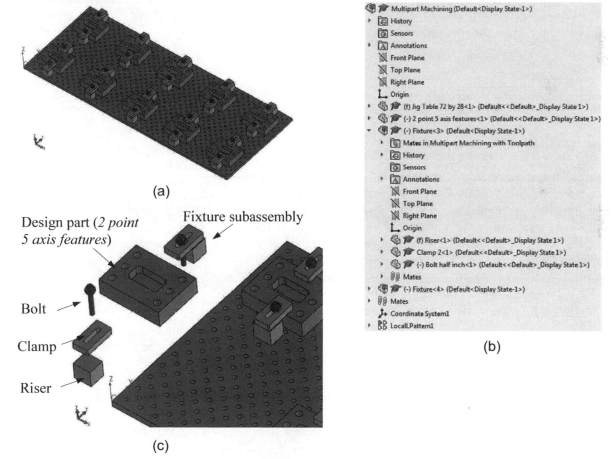

Figure 5.3 The multipart machining assembly example; (a) entire assembly, (b) the solid model tree, and (c) explode and zoom-in view

The part/stock is mounted on the jig table by two fixtures, one on each side. The jig table is 72in.×28in.×0.75in., with threaded holes of 0.5in. in diameter and 0.5in. in depth. The holes are 1.5in. apart every other row and column; that is, along the X and Y directions, respectively, of the coordinate system, *Coordinate System1*, located at the front left corner of the jig table, as shown in Figure 5.3(a) and

Figure 5.3(c). The *2 and 5 axis features* are the only parts (as in ten instances) that will be machined. The fixtures are created as subassemblies; each consists of three parts, riser, clamp, and a bolt, as seen in Figure 5.3(c). In this lesson, we choose *Coordinate System1* for both fixture coordinate system and part setup origin. As a result, the G-code generated refers to this coordinate system. The design part (*2 and 5 axis features.SLDPRT*) with the two fixtures are patterned to create an additional nine instances in two rows as a linear pattern feature (*LocalPattern1* under the *FeatureManager* design tree). The unit system chosen is IPS (inch, pound, second). When you open the assembly model *Multipart Machining.SLDASM*, you should see 2 parts and 2 subassemblies, a coordinate system, and a pattern feature, listed in the feature tree like that of Figure 5.3(b).

You may open the example file with toolpath created (filename: *Multipart Machining with toolpath.SLDASM*) to preview toolpaths created for this example. When you open the file, you may expand the *Part Manager* entity, expand *2 point 5 axis features.SLDPRT*, and then the *Instances* to see the ten instances of the part to be machined (*2 and 5 axis features<1>* to *<10>*), as shown in Figure 5.4. Also, you should see the four operations listed under *Setup1* of the CAMWorks operation tree tab (see Figure 5.4). You may click the *Simulate Toolpath* button above the graphics area to review the machining simulation.

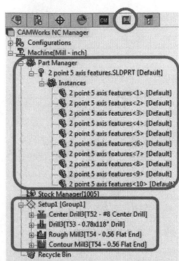

5.3   Using CAMWorks

Open SOLIDWORKS Assembly

Open the assembly model (filename: *Multipart Machining.SLDASM*) downloaded from the publisher's website. This assembly model, as shown in Figure 5.3, consists of four components (jig table, the part: *2 and 5 axis features*, and two fixtures), nine mates, a coordinate system, and a linear pattern. Again, as soon as you open the model, you may want to check the unit system and make sure the IPS system is selected. You may also increase the decimals from the default 2 to 4 digits similar to that of previous lessons.

Figure 5.4 The instances and NC operations listed under the CAMWorks operation tree tab

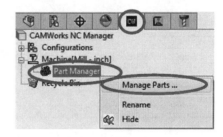

Figure 5.5 Right click *Part Manager* and select *Manage Parts*

Select NC Machine

Click the CAMWorks feature tree tab and right click *Mill-in* to select *Edit Definition*. Similar to those of previous lessons, in the *Machine* dialog box, we select *Mill-in* (already selected by default) under *Machine* tab, choose *Crib 1* under *Available tool cribs* of the *Tool Crib* tab, select *M3AXIS-TUTORIAL* under the *Post Processor* tab, and select *Coordinate System1* under *Fixture Coordinate System* of the *Setup* tab.

Manage Part

Since we are dealing with an assembly with multiple components, we have to identify which part or parts to cut. Under the CAMWorks feature tree tab , right click *Part Manager* and select *Manage Parts* (see Figure 5.5). The *Manage Parts* dialog box appears.

Pick the part (*2 point 5 axis features*) in the graphics area or select *2 point 5 axis features* under the *FeatureManager* design tree tab ⬦. The part is now listed under *Selected Parts* in the *Manage Parts* dialog box (Figure 5.6). Click *Add All Instances* button to bring in all instances, keep the *Subprogram* under *G-code format* selected, and then click *OK* to accept the definition. Selecting the *Subprogram* under *G-code format* will output G-codes with subprogram calls in machining all the instances. Click *OK* to accept the part definition.

Click the CAMWorks feature tree tab ⬛, expand the *Part Manager* entity to see the part (*2 point 5 axis features.SLDPRT*) and its instances like that of Figure 5.7. Also, a *Stock Manager* entity is added to the feature tree, as circled in Figure 5.7.

Figure 5.6 The *Manage Parts* dialog box

Create Stock

From the CAMWorks feature tree ⬛, right click *Stock Manager* and choose *Edit Definition*. In the *Stock Manager* dialog box (Figure 5.8), we leave the default stock size (8in.×6in.×2in.) and material (*Steel 1005*), and click the *Apply Current Definitions to All Parts* button ⬛ (circled in Figure 5.8). Then click the checkmark ✔ to accept the stock definition.

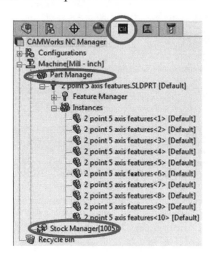

Figure 5.7 Entities listed under the CAMWorks feature tree tab

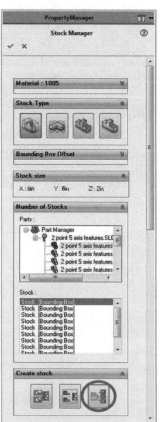

Figure 5.8 The *Stock Manager* dialog box

Extract Machinable Features

Click the *Extract Machinable Features* button 🔘 above the graphics area. A *Setup* entity is created with two machinable features extracted: *Hole Group* (with six holes) and *Rectangular Pocket*, all listed in CAMWorks feature tree ⬛ (Figure 5.9). Both machinable features are shown in magenta color.

In the graphics area (see Figure 5.10), a symbol ⊕ with arrow pointing downward appears at the front left corner of the jig table coinciding with *Coordinate System1* (not shown in Figure 5.10). This symbol indicates that the tool axis is chosen correctly (pointing downward). This corner point also serves as the part setup origin, defining G-code program zero location.

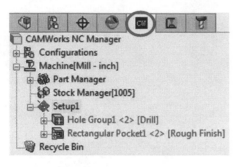

Figure 5.9 The two machinable
features extracted

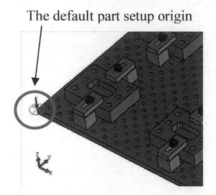

The default part setup origin

Figure 5.10 The default part setup origin
coinciding with *Coordinate System1*

Generate Operation Plan and Toolpath

Click the *Generate Operation Plan* button above the graphics area. Four operations, center drill, drill, rough (cutting the pocket), and contour (cutting the pocket), are generated. They are listed in CAMWorks operation tree (Figure 5.11). Again they are shown in magenta color. Click the *Generate Toolpath* button above the graphics area to create the toolpath. The four operations are turned into black color after toolpaths are generated.

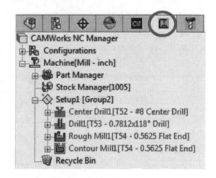

Figure 5.11 The four
operations generated

Part Setup Origin

If you click *Setup1* under CAMWorks operation tree ▣, the part setup origin appears at the front left corner of the jig table, coinciding with the fixture coordinate system by default. We review the options that define the default setup origin.

Click the CAMWorks operation tree tab ▣, and right click *Setup1* to choose *Edit Definition*. In the *Setup Parameters* dialog box, click the *Origin* tab (Figure 5.12). Note that the *Setup origin* is selected for *Output origin*, and the *Fixture coordinate system* is chosen for *Setup origin*. This implies that the fixture coordinate system is chosen as the output origin for G-codes.

We stay with the default selections. We review the G-codes output by CAMWorks based on the selections in Section 5.5. We will explore other options in the exercise problems at the end of the lesson.

Fixtures and Components to Avoid

We will select the jig table and the two fixtures to avoid in generating the toolpath.

In the *Setup Parameters* dialog box, click the *Fixtures* tab (Figure 5.13). Choose the components to avoid, including the jig table and the two fixture subassemblies, either from the graphics area or the

FeatureManager design tree tab  (see Figure 5.14). Click *Add All Instances* and choose *Avoid All*, then click *OK*.

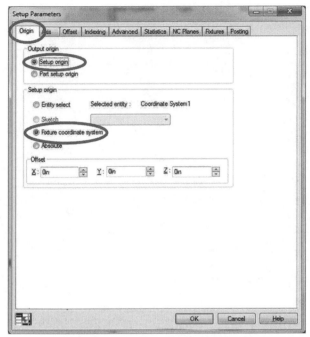

Figure 5.12 The *Origin* tab in the
Setup Parameters dialog box

Figure 5.13 The *Fixtures* tab in the
Setup Parameters dialog box

Click *Yes* to the question in the warning box: *The origin or machining direction or advanced parameters has changed, toolpaths need to be recalculated. Regenerate toolpaths now?* The toolpath will be regenerated like that shown in Figure 5.15.

5.4   The Sequence of Part Machining

You may click the *Simulate Toolpath* button to run the material removal simulation like that of Figure 5.1.

Note that the machining sequence follows a counterclockwise order looking down from above the jig table, from parts labeled 1 close to the front left corner, to 10 close to the rear left corner, as shown in Figure 5.16. The center drill operation ran over all ten instances first, followed by the drill operations. The rough and contour pocket milling operations were carried out together for individual instances 1 to 10.

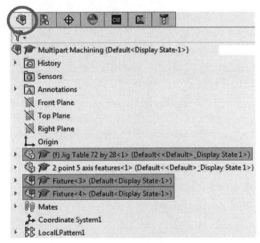

Figure 5.14 Selecting components to avoid

5.5   The G-Codes

In this example, we chose the part setup origin at the front left corner of the top face of the jig table (as shown in Figure 5.10), coinciding with that of the fixture coordinate system, *Coordinate System1*. We expect that the G-codes generated by CAMWorks refer to the part setup origin (that is, the fixture coordinate system, *Coordinate System1*). In addition, we chose *Subprogram* for the *G-code format* in the *Manage Parts* dialog box shown in Figure 5.6. We expect that the G-codes generated by CAMWorks includes subprograms that are called by the main program to perform the four machining operations ten times for the ten respective instances following the order shown in Figure 5.16.

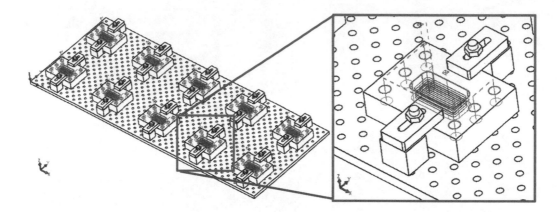

Figure 5.15 Toolpath of the multipart machining example

You may click the *Post Process* button ![Post Process] above the graphics area, and follow the same steps learned in Lesson 2 to convert the toolpaths into G-codes. Figure 5.17 and Figure 5.18 show (partial) contents of the main program (O0001) and subprograms (O0002-O0005), respectively. The four subprograms, O0002 to O0005, carry out center drill, drill, and rough and contour pocket milling, respectively.

Note that the coordinates of the top left corner at the top face of stock 1 is (2, 3.5, 2), stock 2 is (17, 3.5, 2), and stock 10 is (2, 18.5, 2) referring back to *Coordinate System1*, as shown in Figure 5.16. The coordinates of the characteristic points can be found by using the measure capability of SOLIDWORKS (by choosing from the pull-down menu *Tools > Evaluate > Measure*). Therefore, the distance between neighboring parts is 15in. along both X- and Y-directions, as shown in Figure 5.16. Since the size of the block is 8in.×6in.×2in., these are the centers of the individual stocks: stock 1 (6, 6.5, 2), stock 2 (21, 6.5, 2), stock 3 (36, 6.5, 2), stock 4 (51, 6.5, 2), stock 5 (66, 6.5, 2), stock 6 (66, 21.5, 2), stock 7 (51, 21.5, 2), stock 8 (36, 21.5, 2), stock 9 (21, 21.5, 2), and stock 10 (6, 21.5, 2).

As shown in Figure 5.17, the main program is written in three segments, center drill operations defined in the NC blocks N1 to N46 (first column), drill operations defined in blocks N47 to N91 (second column), and rough and contour mill operations defined in N92 to N179 (third column). In the first segment, a local coordinate system was assigned at the center point of the top face of individual stocks using the G52 NC word; for example, blocks N8 (stock 1), N12 (stock 2), and N44 (stock 10). Subprogram O0002 was called using M98 NC word right after the local coordinate system was set; for example, blocks N9 (stock 1), N13 (stock 2), and N45 (stock 10). G52 appears again in the main program right after returning from subprogram call to set the coordinate system back to (0,0,0).

A complete content of subprogram O0002 is shown in the first column of Figure 5.18. The center locations of the six holes are specified in blocks N1 and N4-N8 referring to the center point of the top face of the respective stock, in which the local coordinate system is located. Note that since the part setup origin was set at the top left corner of the jig table, the main program sets local coordinate system at the center point of the top face of the stock by referring back to the part setup origin, which is perfectly fine. This is how the post process was written.

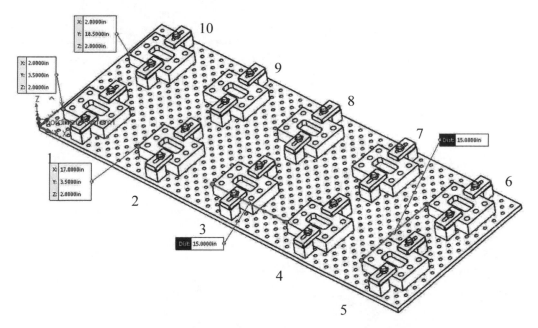

Figure 5.16 Locations of key corner points and distances between parts

O0001	N47 G00 G91 G28 Z0	N92 G91 G28 Z0
N1 G20	N48 (25/32 JOBBER DRILL)	N93 (9/16 EM CRB 4FL 1-1/8 LOC)
N2 (#8 HSS 60Deg Centerdrill)	N49 T53 M06	N94 T54 M06
N3 G91 G28 X0 Y0 Z0	N50 S1000 M03	N95 S200 M03
N4 T52 M06		
N5 S2800 M03	N51 (Drill1)	N96 (Rough Mill1)
	N52 G90	N97 G90
N6 (Center Drill1)	N53 G52 X6. Y6.5 Z2.	N98 G52 X6. Y6.5 Z2.
N7 G90	N54 M98 P0003	N99 M98 P0004
N8 G52 X6. Y6.5 Z2.	N55 G52 X0 Y0 Z0	N100 G52 X0 Y0 Z0
N9 M98 P0002		
N10 G52 X0 Y0 Z0	N56 (Drill1)	N101 (Contour Mill1)
	N57 G52 X21. Y6.5 Z2.	N102 G52 X6. Y6.5 Z2.
N11 (Center Drill1)	N58 M98 P0003	N103 M98 P0005
N12 G52 X21. Y6.5 Z2.	N59 G52 X0 Y0 Z0	N104 G52 X0 Y0 Z0
N13 M98 P0002		
N14 G52 X0 Y0 Z0	N60 (Drill1)	N105 (Rough Mill1)
	N61 G52 X36. Y6.5 Z2.	N106 G52 X21. Y6.5 Z2.
N15 (Center Drill1)	N62 M98 P0003	N107 M98 P0004
N16 G52 X36. Y6.5 Z2.	N63 G52 X0 Y0 Z0	N108 G52 X0 Y0 Z0
N17 M98 P0002		
N18 G52 X0 Y0 Z0		N109 (Contour Mill1)
	...	N110 G52 X21. Y6.5 Z2.
		N111 M98 P0005
...	N88 (Drill1)	N112 G52 X0 Y0 Z0
	N89 G52 X6. Y21.5 Z2.	
N43 (Center Drill1)	N90 M98 P0003	
N44 G52 X6. Y21.5 Z2.	N91 G52 X0 Y0 Z0	...
N45 M98 P0002		
N46 G52 X0 Y0 Z0		N173 (Contour Mill1)
		N174 G52 X6. Y21.5 Z2.
		N175 M98 P0005
		N176 G52 X0 Y0 Z0
		N177 G91 G28 Z0
		N178 G28 X0 Y0
		N179 M30

Figure 5.17 The G-code, main program

```
O0002                          O0003                          O0004                              O0005
N1 G90 G54 G00 X-3. Y2.        N1 G90 G54 G00 X-3. Y2.        N1 G90 G54 G00 X-1.1463 Y.1463      N1 G90 G54 G41 D74 G00 X0 Y.7188
N2 G43 Z1. H52 M08             N2 G43 Z1. H53 M08             N2 G43 Z5. H54 M08                  N2 G43 Z1. H54 M08
N3 G81 G98 R.9469 Z-.6914 F6.  N3 Z.9469                      N3 Z1.                              N3 Z.1
N4 Y0                          N4 G81 G99 R.9469 Z-1. F10.    N4 G01 Z-.225 F1.25                 N4 G01 Z-.225 F3.75
N5 Y-2.                        N5 Y0                          N5 G17 Y-.1462 F10.                 N5 G17 X-1.5 F30.
N6 X3.                         N6 Y-2.                        N6 X1.1463                          N6 G03 X-1.7188 Y.5 I0 J-.2188
N7 Y0                          N7 X3.                         N7 Y.1463                           N7 G01 Y-.5
N8 Y2.                         N8 Y0                          N8 X-1.1463                         N8 G03 X-1.5 Y-.7188 I.2188 J0
N9 G80 Z1. M09                 N9 Y2.                         N9 Y.4275                           N9 G01 X1.5
N10 M99                        N10 G80 Z1. M09                N10 X-1.4275                        N10 G03 X1.7187 Y-.5 I0 J.2188
                               N11 M99                        N11 Y-.4275                         N11 G01 Y.5
                                                              N12 X1.4275                         N12 G03 X1.5 Y.7188 I-.2187 J0
                                                              N13 Y.4275
                                                              N14 X-1.1463                        ...
                                                              N15 Y.7088
                                                              N16 X-1.5
                                                                                                  N60 G03 X1.5 Y.7188 I-.2187 J0
                                                                                                  N61 G01 X0
                                                              ...                                 N62 G00 Z.1
                                                                                                  N63 Z1. M09
                                                              N164 G03 X1.5 Y.7088 I-.2087 J0     N64 G40 X0 Y.7188
                                                              N165 G01 X-1.1463                    N65 M99
                                                              N166 G00 Z1.
                                                              N167 Z5. M09
                                                              N168 M99
```

Figure 5.18 The G-codes, subprograms, O0002, O0003, O0004, and O0005

Similarly for drill operations, the main program (second column in Figure 5.17) sets local coordinate systems (G52) and calls subprogram O0003 (M98). The contents of subprogram O0003 shown in the second column of Figure 5.18 are similar to those of subprogram O0002.

For the pocket milling operations, as shown in the third column of Figure 5.17, the main program sets local coordinate systems (G52) and calls subprogram O0004 for rough milling and then O0005 for contour milling (M98). The partial contents of subprograms O0004 and O0005 are shown in the third and fourth columns of Figure 5.18, respectively. The X- and Y-coordinates of the cutter locations shown in the subprograms O0004 and O0005 are referred to the respective local coordinate systems, which are again the center point at the top face of the respective stocks, set by using G52 in the main program.

We have verified that G-codes are correctly generated. We have now completed this exercise. You may save your model for future reference.

5.6   Exercises

Problem 5.1.   Create an assembly like that of Figure 5.19 using the parts (*Jig table*, *Problem 5.1 Part*, *Clamp*, and *Shot Bolt*) and subassembly (*Fixture*) in the Problem 5.1 folder downloaded from the publisher's website. Note that *Problem 5.1 Part* is identical to that of Problem 4.1. Create a total of eight instances as a linear pattern feature.

(a)   Generate a machining simulation for the eight instances using the machining operations created in Problem 4.1.

(b)   Generate G-codes with default selections like those discussed in this lesson. Verify that the codes are generated correctly by reviewing the contents of the codes, similar to those of Section 5.5.

Hint: You may perform the following to manually create a machinable feature in assembly, (a) create instances by right clicking *Part Manager* and choosing *Manage Parts*, (b) insert a mill part setup by expanding *Problem 5.1 Part* and right clicking *Feature Manager*, and (c) insert a new multi surface feature by expanding *Feature Manager* and right clicking *Mill Part Setup*.

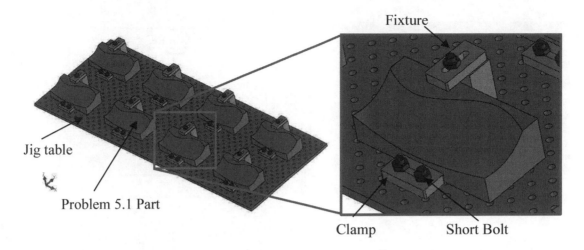

Figure 5.19 The assembly model of Problem 5.1

Problem 5.2. Continue from Problem 5.1. In this case, we deselect the *Subprogram* under *G-code format* in the *Manage Parts* dialog box (see Figure 5.20) and output G-codes. Open the G-codes and identify the differences that this selection makes to the G-codes ouput. Verify if the G-codes are output correctly.

Figure 5.20 The *Manage Parts* dialog box

Problem 5.3. Now we get back to the example of this lesson to explore other options in selecting part setup origin. Bring out the *Setup Parameters* dialog box and select *Part setup origin* for *Output origin* (see Figure 5.21). Note that the part setup origin symbols appear at the front left corner at the top face of individual stocks, as shown in Figure 5.22.

Output G-codes. Open the G-code file and identify the differences that this selction makes to the G-codes ouput in Section 5.5. Verify if the G-codes are output correctly.

Figure 5.21 The *Origin* tab in the
Setup Parameters dialog box

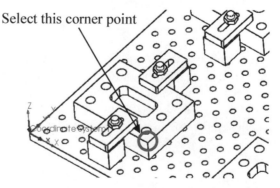

The part setup
origin symbols

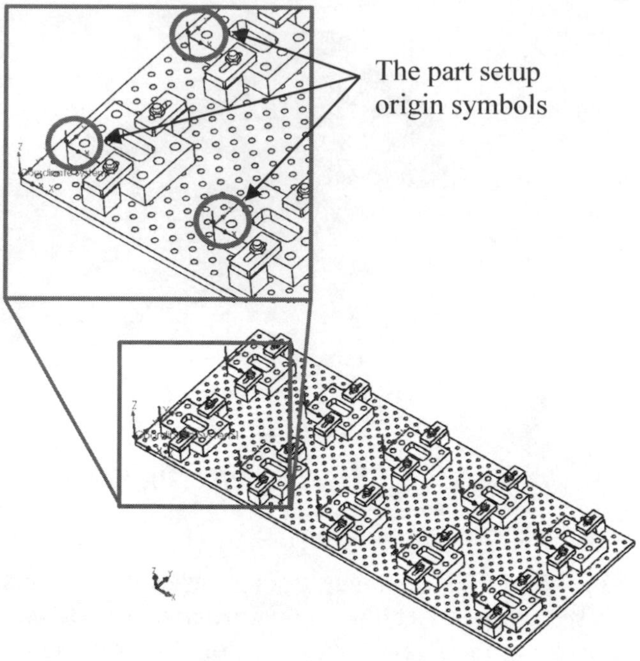

Figure 5.22 The part setup origin symbols

Problem 5.4. Continue from Problem 5.3. In this case, we choose *Setup origin* for *Output origin* (see Figure 5.23), click *Entity select* under *Setup origin*, and pick a corner point shown in Figure 5.24. Then, we output G-codes. Open the G-codes and identify the differences that such options make to the G-codes ouput in Problem 5.3. Verify if the G-codes are output correctly.

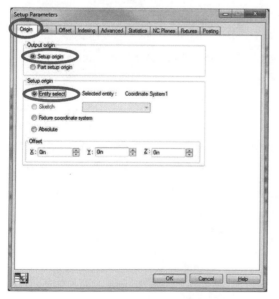

Figure 5.23 The *Origin* tab in the
Setup Parameters dialog box

Select this corner point

Figure 5.24 Select a corner point at part
setup origin

Lesson 6: Multiplane Machining

6.1 Overview of the Lesson

In this lesson, we will learn to create machining operations that cut parts with machinable features on multiple planes. In particular, we assume the stocks to be mounted on a tombstone that is rotated with desired angles by a rotary table rotating in a direction of the so-called 4th axis (or A-axis).

The design part is similar to that of Lessons 3 and 5, except that there are additional holes on the side faces, in addition to a pocket and holes on the top face. Similar to Lesson 5, the face milling discussed in Lesson 3 is not included to simplify the fixture design for stock setup. The part is assembled to a tombstone by using two fixtures (one on each end). Each fixture consists of a clamp and two bolts. A total of four parts mounted on the respective four faces of the tombstone are to be machined, as shown in Figure 6.1. The tombstone is mounted on a rotary table using four bolts, and the rotary table is assembled to a rotary unit, which is mounted on a jig table like that of Lesson 5 (not shown in Figure 6.1). We assume that a 3-axis mill with a rotary table is employed for this lesson. The rotary table rotates ±90° or ±180° along the longitudal direction for the tools to cut features on faces that are normal to the tool axis.

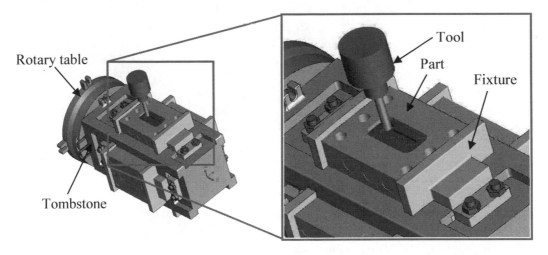

Figure 6.1 Material removal simulation of the multiplane machining example

We will learn the steps to define the 4th axis, in addition to creating instances of part for generating machining operations that are similar to that of Lesson 5, including defining stocks for individual instances, extracting machinable features, generating toolpath, and selecting components in the assembly for the tools to avoid. We will take a closer look at the G-codes generated by CAMWorks to verify that the codes are generated correctly to support machine operations for a 3-axis mill with a rotary table.

6.2   The Multiplane Machining Example

A design part (filename: *2 point 5 axis features with side holes.SLDPRT*) similar to that of Lessons 3 and 5 with a bounding box of 8in.×6in.×2in. is employed in this lesson. In addition to the pocket and holes (three at each end) on its top face—see Figure 6.2(a)—there are three holes on its front side face, as shown in Figure 6.2(a), and another three holes on its rear side face; see Figure 6.2(b). All these features are of the same size, and they are extracted as machinable features by using the automatic feature recognition (AFR) capability.

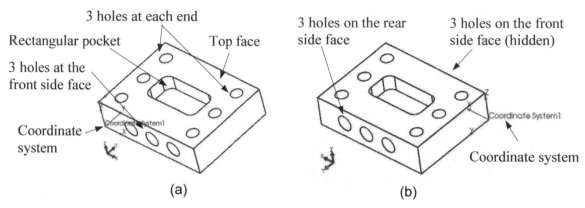

(a)   (b)

Figure 6.2 Machinable features of the design part, *2 and 5 axis features with side holes*, (a) features at the top and front side faces, and (b) holes on the rear side face (a rotated view)

The assembly model (*Multiplane Machining.SLDASM*) consists of eight parts, one subassembly (called *clamped part*), a circular pattern feature that includes the instances of the clamped part subassembly, a point and a coordinate system, as listed in the *FeatureManage* design tree 🔳 shown in Figure 6.3(a).

The jig table of Lesson 5 is employed again for this lesson. A rotary unit is mounted on the jig table; see Figure 6.3(b). And a rotary table is assembled to the rotary unit by using four 0.5in. bolts, and a tombstone on which parts are mounted is assembled to the rotary table. The clamped part subassembly consisting of the design part (*2 and 5 axis features with side holes*), two fixtures, and two bolts are mounted on the respective four faces of the tombstone; see Figure 6.3(c). A coordinate system, *Coordinate System1*, is defined at the center of the front end face of the tombstone with X-axis pointing along the longitudinal direction of the jig table; see Figure 6.3(b). Note that *Coordinate System1* will be chosen as the fixture coordinate system for the machining opertations in this lesson. The rotary table provides rotation motion along the X-axis of the fixture coordinate system.

There are four setups corresponding to the respective four tool axes defined in this example. As shown in Figure 6.4(a), the tool axis (symbol: 🔧) of *Setup1* points in the –Z direction with four machinable features: *Hole Group1* and *Rectangular Pocket1* on the top face of the part mounted on the top face of the tombstone, *Hole Group2* on the side face of the part mounted on the front side face of the tombstone, and *Hole Group3* on the part mounted on the rear side face. As another example, the tool axis of *Setup2* points in the +Y direction—see Figure 6.4(b)—with another four machinable features, *Hole Group1* and *Rectangular Pocket1* on the part mounted on the front side face of the tombstone, *Hole Group2* on the part mounted on the bottom face, and *Hole Group3* on the part of the top face of the tombstone. Tool axes of *Setup3* and *Setup4* point in –Y and +Z directions, respectively. Machinable features associated with these two setups are pockets and holes sketched on the part faces that are normal to the respective tool axes, similar to those of *Setup1* and *Setup2* shown in Figure 6.4.

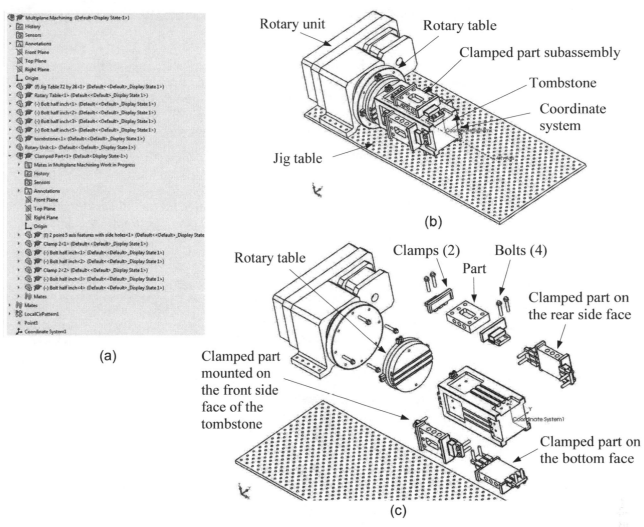

(a)

(b)

(c)

Figure 6.3 The solid model of multiplane machining assembly example: (a) model tree, (b) entire assembly, and (c) explode view

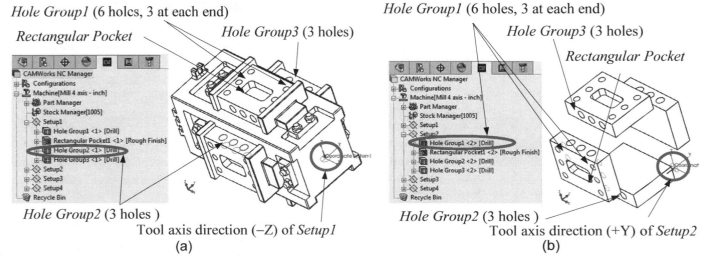

Hole Group1 (6 holcs, 3 at each end)

Rectangular Pocket

Hole Group3 (3 holes)

Hole Group2 (3 holes)

Tool axis direction (–Z) of *Setup1*

(a)

Hole Group1 (6 holes, 3 at each end)

Hole Group3 (3 holes)

Rectangular Pocket

Hole Group2 (3 holes)

Tool axis direction (+Y) of *Setup2*

(b)

Figure 6.4 The setups generated for the multiplane machining example: (a) *Setup1* and associated machinable features, and (b) *Setup2* and associated machinable features (tombstone and fixtures not shown)

Note that you might see the four setups in a different order as in the example file when you go through this lesson.

The stock size is identical to that of the bounding box of the part. The pocket milling operations, including a rough milling and a contour milling, and hole drilling operations are identical to those of Lesson 5 (for example, the first four operations of *Setup1* shown in Figure 6.5). Similarly, the drilling operations for holes of parts mounted on the side faces of the tombstone consist of a center drill and a drill operation (see Figure 6.5). The toolpaths of *Setup1* is shown in Figure 6.5 for illustration. Similar toolpaths can be found for the other three setups. Note that the part setup origin coincides with that of the origin of *Coordinate System1*, as pointed out in Figure 6.5.

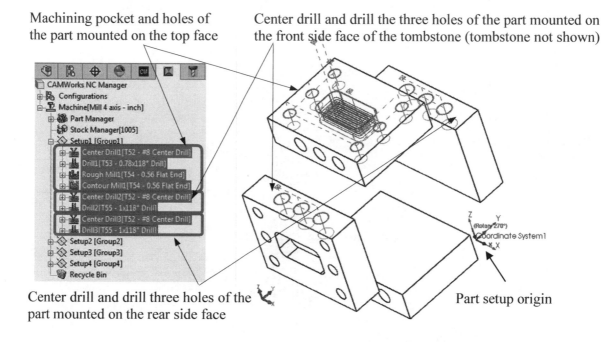

Machining pocket and holes of the part mounted on the top face

Center drill and drill the three holes of the part mounted on the front side face of the tombstone (tombstone not shown)

Center drill and drill three holes of the part mounted on the rear side face

Part setup origin

Figure 6.5 Toolpaths of the eight operations of *Setup1* (tombstone and fixtures not shown)

You may open the example file with toolpaths created (filename: *Multiplane Machining with toolpath.SLDASM*) to preview the toolpaths of this example. When you open the file, you may expand the *Part Manager* entity, expand the part (*2 point 5 axis features with side holes.SLDPRT*), and then expand the *Instances* to see the four instances of the part to be cut (*Clamped part-1 to 4* shown in Figure 6.6). Also, you should see the eight operations listed under *Setup1* of the CAMWorks operation tree tab, as shown in Figure 6.6. Similar operations can be seen for the other three setups simply by expanding the respective entities in the operation tree. You may click the *Simulate Toolpath* button above the graphics area to preview the machining operations.

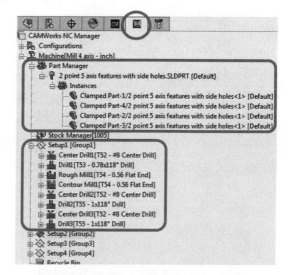

Figure 6.6 The instances and NC operations listed under the CAMWorks operation tree tab

6.3　Using CAMWorks

Open SOLIDWORKS Assembly

Open the assembly model (filename: *Multiplane Machining.SLDASM*) downloaded from the publisher's website. This assembly model appears in the graphics area similar to that of Figure 6.3(b). Again, as soon as you open the model, you may want to check the unit system chosen and make sure the IPS system is selected. You may also increase the decimals from the defaut 2 to 4 digits similar to that of previous lessons.

Select NC Machine

Click the CAMWorks feature tree tab ▣ and right click *Mill-in* to select *Edit Definition*. In the *Machine* dialog box, we choose *Mill 4 axis-inch* under *Machine* tab (Figure 6.7(a)) and click *Select*, choose *Crib 1* under *Available tool cribs* of the *Tool Crib* tab, select *M4AXIS-TUTORIAL* under the *Post Processor* tab (Figure 6.7(b)), and select *Coordinate System1* under *Fixture Coordinate* system of the *Setup* tab.

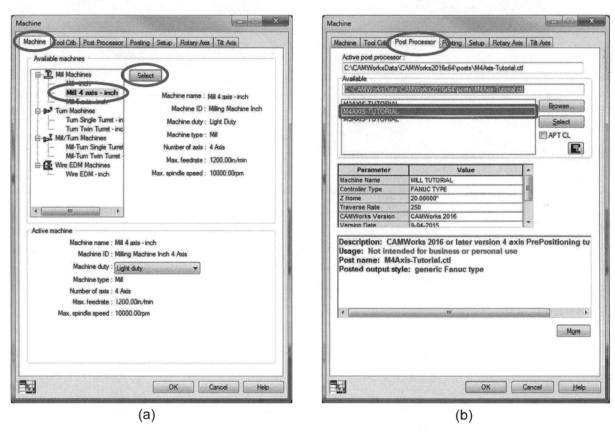

(a)　　　　　　　　　　　　　　　　　　(b)

Figure 6.7　The *Machine* dialog box, (a) selecting 4 axis mill, and (b) selecting the 4 axis post-processor

Define the Rotary Axis

In the *Machine* dialog box, click the *Rotary Axis* tab, choose *X axis* as the *Rotary axis* and *XZ plane* as the *0 degree position* [circled in Figure 6.8(a)]. Then, click *OK*. In the graphics area, a circular arc with a counterclockwise arrow appears at the origin of *Coordinate System1*, as shown in Figure 6.8(b), indicating the rotation direction of the rotary axis.

Manage Part

Similar to Lesson 5, we are dealing with an assembly of multiple components. Therefore, we have to identify which part or parts to cut. Under the CAMWorks feature tree tab ▣ , right click *Part Manager* and select *Manage Parts* (Figure 6.9). The *Manage Parts* dialog box appears.

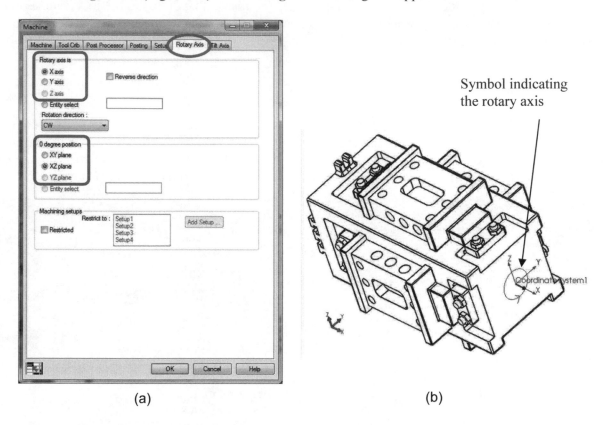

(a)   (b)

Figure 6.8 Defining the rotary axis, (a) the *Rotary Axis* tab of the *Machine* dialog box, and (b) the symbol indicating the rotary axis

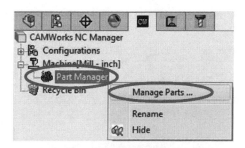

Figure 6.9 Right click *Part Manager* and select *Manage Parts*

Figure 6.10 The *Manage Parts* dialog box

Pick the design part (*2 point 5 axis features with side holes*) in the graphics area or expand *Clamped Part*, and select *2 point 5 axis features with side holes* under the *FeatureManager* design tree tab 🔧. The part is now listed under *Selected Parts* in the *Manage Parts* dialog box (Figure 6.10). Click *Add All Instances* button to bring in all instances, do not select the *Subprogram* under *G-code format* option, and then click *OK* to accept the definition. Not selecting the *Subprogram* under *G-code format* option will output G-codes without subprogram calls. Click *OK* to accept the part definition.

Click the CAMWorks feature tree tab 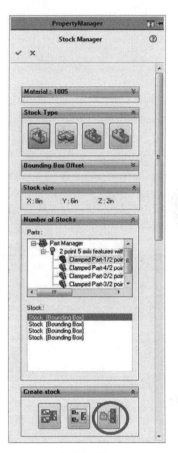, expand the *Part Manager* entity to see the part (*2 point 5 axis features with side holes.SLDPRT*) and its instances to make sure that all four parts are included, as shown in Figure 6.11. Also, a *Stock Manager* entity has been added.

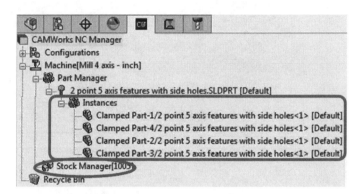

Figure 6.11 All four instances of the part listed under CAMWorks feature tree

Figure 6.12 The *Stock Manager* dialog box

Create Stock

From CAMWorks feature tree 🔲, right click *Stock Manager* and choose *Edit Definition*. In the *Stock Manager* dialog box (Figure 6.12), we leave the default stock size (8in.×6in.×2in.) and material (*Steel 1005*), and click the *Apply Current Definitions to All Parts* button 📋 (circled in Figure 6.12). Then click the checkmark ✔ to accept the definition of stock.

Extract Machinable Features

Click the *Extract Machinable Features* button 📋 above the graphics area. Four setup entities (*Setup1-4*) are created with the respective four machinable features extracted: *Hole Group1* (with six holes), *Rectangular Pocket1, Hole Group2* (with three holes), and *Hole Group3* (with three holes), listed in CAMWorks feature tree 🔲 (Figure 6.13). All machinable features are shown in magenta color.

Click a setup to locate its part setup origin and tool axis in the graphics area. For example, click *Setup4* to see a symbol 🔧 with arrow facing upward (coinciding with the Z-axis) appearing at the center of the front end face of the tombstone (coinciding with *Coordinate System1*, as shown in Figure 6.14). All four setups shared the same part setup point but with different tool axes. Tool axes of the four setups are pointing respectively in –Z, +Y, –Y, and +Z directions.

Generate Operation Plan and Toolpath

Click the *Generate Operation Plan* button ▦ above the graphics area. Eight operations, *Center Drill1*, *Drill1*, *Rough Mill1*, *Contour Mill1*, *Center Drill2*, *Drill2*, *Center Drill3*, and *Drill3*, are generated for each setup. There is a total of 32 operations, eight per setup. They are listed in CAMWorks operation tree ▦ (see Figure 6.15, showing operations of *Setup1* and *Setup4*). Again they are shown in magenta color. Click the *Generate Toolpath* button ▦ above the graphics area to create the toolpath. The operations are turned into black color after toolpaths are generated.

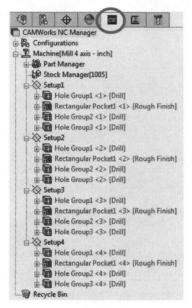

Figure 6.13 The machinable
features extracted

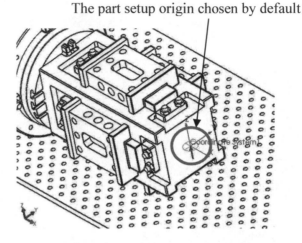

The part setup origin chosen by default

Figure 6.14 The default part setup origin
coinciding with *Coordinate System1*

Fixtures and Components to Avoid

Similar to Lesson 5, we will select components to avoid in generating the toolpath. In this lesson, we select the rotary table, tombstone, fixtures, and bolts to avoid. We will not select the rotary unit and the jig table since the chance that the tools collide with them is minimum for a 3-axis mill with a rotary table.

On the other hand, only the components selected as fixtures are included in the material removal simulation. Excluding the jig table and the rotary unit makes the simulation more visually logical and realistic.

Click the CAMWorks operation tree tab ▦, and right click *Setup1* to choose *Edit Definition*. In the *Setup Parameters* dialog box, click the *Features* tab (Figure 6.16). Choose the components to be part of the fixtures, including the rotary table, tombstone, both clamps, and bolts, either from the graphics area or from the *FeatureManager* design tree tab ▦ (see Figure 6.17) by clicking them.

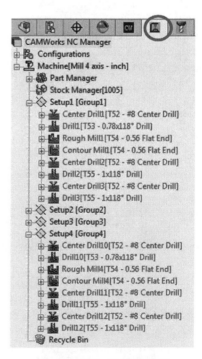

Figure 6.15 The eight
operations generated for
each setup

Click both clamps and bolts from the remaining three sets, including those of the circular pattern feature of the solid model. You may expand pattern feature (*LocalCirPattern1* under *FeatureManager* design tree) to select the clamps and bolts. After selecting all these components, we click the *Avoid All* button to avoid them in the toolpath generation, then we click *OK*.

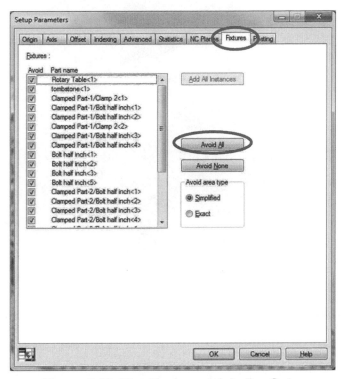

Figure 6.16 The *Features* tab in the *Setup Parameters* dialog box

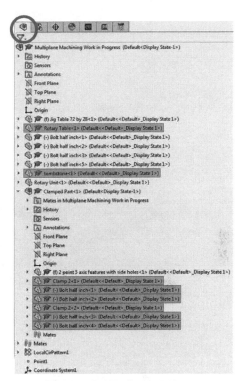

Figure 6.17 Select the components to avoid

Click *Yes* to the question in the warning box: *The origin or machining direction or advanced parameters has changed, toolpaths need to be recalculated. Regenerate toolpaths now?* The toolpaths will be regenerated, for example, like that shown in Figure 6.5 for *Setup1*.

6.4 The Sequence of Part Machining

You may click the *Simulate Toolpath* button to run the material removal simulation like that of Figure 6.1.

Note that the machining sequence follows that of the four setups, from *Setup1* to *Setup4*. The rotary table rotates along the X-axis of *Coordinate System1*. The eight machining operations of *Setup1* cut machinable features on parts labelled 1, 2, and 4, as shown in Figure 6.18(a). The first two operations (center drill and hole drilling) drill the six holes on the top face of part 1, followed by two operations that cut the pocket (rough and contour mills) on part 1. Then two operations drill the three holes (center drill and hole drilling) on part 4. The final two operations cut the three holes on part 2. The eight operations repeat three more times for the remaining respective three setups, as shown in Figure 6.18(b), (c), and (d).

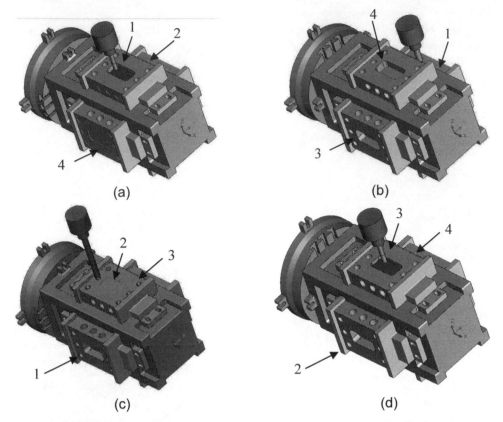

Figure 6.18 Material removal simulation (stock rotates), (a) *Setup1*, (b) *Setup2*, (c) *Setup3*, and (d) *Setup4*

Figure 6.19 Click the *Options* button in the *Toolpath Simulation* tool box

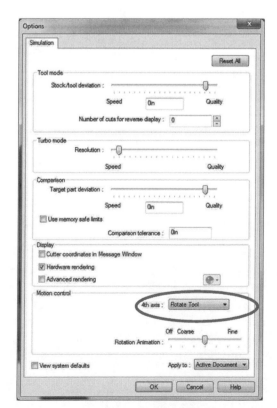

In the default setting, stock rotates in the material removal simulation. The rotary table rotates a –90° angle from *Setup1* to *Setup2*, another 180° angle from *Setup2* to *Setup3*, then a 90° angle from *Setup3* to *Setup4*, and finally a –180° angle to return to *Setup1*.

You may change the default setting to rotate the tool in the material removal simulation (which is less desirable).

This can be done by clicking the *Options* button in the *Toolpath Simulation* tool box (Figure 6.19). In the *Options* dialog box (Figure 6.20), choose *Rotate Tool* for *4th axis* under *Motion control*, and then click *OK*.

Figure 6.20 The *Features* tab in the *Setup Parameters* dialog box

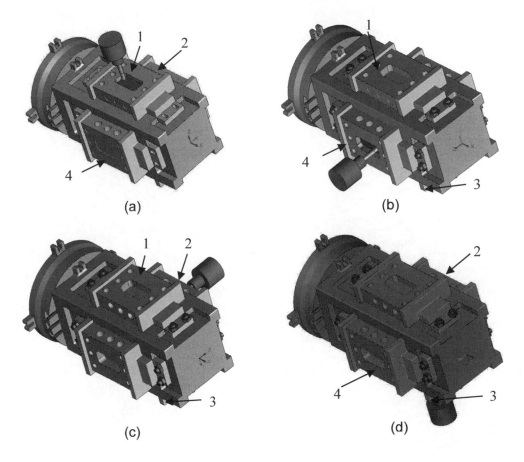

Figure 6.21 Material removal simulation (tool rotates), (a) *Setup1*, (b) *Setup2*, (c) *Setup3*, and (d) *Setup4*

Click *Yes* to the message: *Simulation must be restarted before new settings will take effect. Do you want to restart simulation?*

Click the *Run* button in the *Toolpath Simulation* tool box. The material removal simulation will take place for all four setups, in which the tools rotate like those of Figure 6.21.

6.5 Reviewing The G-Codes

In this example, we chose the part setup origin at the center point of the front face of the tombstone, coinciding with that of the fixture coordinate system, *Coordinate System1*, as shown in Figure 6.4. We expect that the G-code generated by CAMWorks refers to the part setup origin at the coordinate system, *Coordinate System1*. In addition, recall that we did not choose *Subprogram* for the *G-code format* in the *Manage Parts* dialog box (see Figure 6.10). We expect that the G-code generated does not include subprogram calls. Therefore, the main program performs the eight machining operations four times for the respective four setups following the order shown in Figure 6.18.

To understand the G-code, we first locate the center points of a few selected holes. We use the *Measure* option (*Tools > Evaluate > Measure*) of SOLIDWORKS and choose *Center to Center* option; see Figure 6.22(a). We first select the coordinate system (*Coordinate System1*) and pick the edge of one of the holes of the part mounted on top; see Figure 6.22(b) for the hole selected. The coordinates of the center of the selected hole can be located at (−12.25, −2, 6.5), as shown in Figure 6.22(b). Since the distance between the neighboring holes is 2in., the coordinates of the center points of the remaining two holes at the rear end of the top face are, respectively, (−12.25, 0, 6.5), and (−12.25, 2, 6.5).

Similarly, we select the coordinate system (*Coordinate System1*) and pick the edge of one of the holes of the part mounted on the front side face; see Figure 6.22(c) for the hole picked. The coordinates of the hole center are (−10.75, −5.5, 3.0), as shown in Figure 6.22(c). Since the distance between the neighboring holes is 1.75in., the coordinates of the center points of the remaining two holes are, respectively, (−9.0, −5.5, 3.0), and (−7.25, −5.5, 3.0).

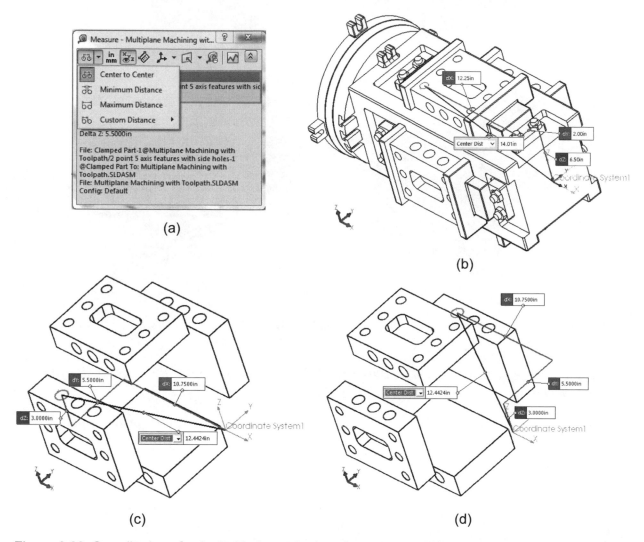

Figure 6.22 Coordinates of selected hole center locations, (a) the *Measure* dialog box, (b) a hole of the part mounted on the top face of the tombstone selected, (c) a hole of the part mounted on the front side face of the tombstone selected (tombstone not shown), and (b) another hole of the part on the rear side face selected (tombstone not shown)

Also, we select the coordinate system (*Coordinate System1*) and pick the edge of the hole of the part mounted on the rear side of the tombstone; see Figure 6.22(d). The coordinates of the hole center are (−10.75, 5.5, 3.0), as shown in Figure 6.22(d). Since the distance between the neighboring holes is 1.75in., the center points of the remaining holes are, respectively, (−9.00, 5.5, 3.0), and (−7.25, 5.5, 3.0).

```
O0001
N1 G20
N2 (#8 HSS 60Deg Centerdrill)
N3 G91 G28 X0 Y0 Z0
N4 T52 M06
N5 S2800 M03

N6 ( Center Drill1 )
N7 G90 G54 G00 X-12.25 Y2. A-90. B0
N8 G43 Z7.5 H52 M08
N9 G81 G98 R6.6 Z5.8086 F6.
N10 Y0
N11 Y-2.
N12 X-6.25
N13 Y0
N14 Y2.
N15 G80 Z7.5 M09
N16 G91 G28 Z0
N17 (25/32 JOBBER DRILL)
N18 T53 M06
N19 S1000 M03

N20 ( Drill1 )
N21 G90 G54 G00 X-12.25 Y2.
N22 G43 Z6.6 H53 M08
N23 G81 G99 R6.6 Z5.5 F10.
N24 Y0
N25 Y-2.
N26 X-6.25
N27 Y0
N28 Y2.

...

N34 ( Rough Mill1 )
N35 G90 G54 G00 X-10.3963 Y.1462
N36 G43 Z7.5 H54 M08
N37 G01 Z6.275 F1.25
N38 Y-.1463 F10.
N39 X-8.1038
...

N201 ( Contour Mill1 )
N202 S2200
N203 X 9.25 Y.7187
N204 Z7.5
N205 Z6.6
N206 G01 Z6.275 F3.75
N207 G41 D74 X-10.75 Y.7187 F30.
...

271 ( Center Drill2 )
N272 G90 G54 G00 X-10.75 Y-5.5
N273 G43 Z4. H52 M08
N274 G81 G98 R3.1 Z2.3086 F6.
N275 X-9
N276 X-7.25
N277 G80 Z4. M09
...

N282 ( Drill2 )
N283 G90 G54 G00 X-10.75 Y-5.5
N284 G43 Z3.1 H55 M08
N285 G81 G99 R3.1 Z1.75 F10.
N286 X-9.
N287 X-7.25
...

N293 ( Center Drill3 )
N294 G90 G54 G00 X-10.75 Y5.5
N295 G43 Z4. H52 M08
N296 G81 G98 R3.1 Z2.3086 F6.
N297 X-9.
N298 X-7.25
...

N315 ( Center Drill4 )
N316 G90 G54 G00 X-12.25 Y2. A-180.
N317 G43 Z7.5 H52 M08
N318 G81 G98 R6.6 Z5.8086 F6.
N319 Y0
N320 Y-2.
```

```
N321 X-6.25
N322 Y0
N323 Y2.
N324 G80 Z7.5 M09
N325 G91 G28 Z0
N326 (25/32 JOBBER DRILL)
N327 T53 M06
N328 S1000 M03

N329 ( Drill4 )
N330 G90 G54 G00 X-12.25 Y2.
N331 G43 Z6.6 H53 M08
N332 G81 G99 R6.6 Z5.5 F10.
N333 Y0
N334 Y-2.
N335 X-6.25
N336 Y0
N337 Y2.
N338 G80 Z7.5 M09
N339 G91 G28 Z0
N340 (9/16 EM CRB 4FL 1-1/8 LOC)
N341 T54 M06
N342 S200 M03

N343 ( Rough Mill2 )
N344 G90 G54 G00 X-10.3963 Y.1462
N345 G43 Z7.5 H54 M08
N346 G01 Z6.275 F1.25
N347 Y-.1463 F10.
N348 X-8.1038

...

N602 ( Center Drill6 )
N603 G90 G54 G00 X-10.75 Y5.5
N604 G43 Z4. H52 M08
N605 G81 G98 R3.1 Z2.3086 F6.
N606 X-9.
N607 X-7.25
N608 G80 Z4. M09
N609 G91 G28 Z0
N610 (1 JOBBER DRILL)
N611 T55 M06
N612 S1000 M03

...

N624 ( Center Drill7 )
N625 G90 G54 G00 X-12.25 Y2. A0
N626 G43 Z7.5 H52 M08
N627 G81 G98 R6.6 Z5.8086 F6.
N628 Y0
N629 Y-2.
N630 X-6.25
N631 Y0
N632 Y2.
N633 G80 Z7.5 M09
N634 G91 G28 Z0
N635 (25/32 JOBBER DRILL)
N636 T53 M06
N637 S1000 M03

N638 ( Drill7 )
N639 G90 G54 G00 X-12.25 Y2.
N640 G43 Z6.6 H53 M08
N641 G81 G99 R6.6 Z5.5 F10.
N642 Y0
N643 Y-2.
N644 X-6.25
N645 Y0
N646 Y2.
N647 G80 Z7.5 M09
N648 G91 G28 Z0
N649 (9/16 EM CRB 4FL 1-1/8 LOC)
N650 T54 M06
N651 S200 M03

N652 ( Rough Mill3 )
N653 G90 G54 G00 X-10.3963 Y.1463
N654 G43 Z7.5 H54 M08
N655 G01 Z6.275 F1.25
N656 Y-.1462 F10.
N657 X-8.1038
N658 Y.1463
...
```

```
N933 ( Center Drill10 )
N934 G90 G54 G00 X-12.25 Y2. A90.
N935 G43 Z7.5 H52 M08
N936 G81 G98 R6.6 Z5.8086 F6.
N937 Y0
N938 Y-2.
N939 X-6.25
N940 Y0
N941 Y2.
N942 G80 Z7.5 M09
N943 G91 G28 Z0
N944 (25/32 JOBBER DRILL)
N945 T53 M06
N946 S1000 M03

N947 ( Drill10 )
N948 G90 G54 G00 X-12.25 Y2.
N949 G43 Z6.6 H53 M08
N950 G81 G99 R6.6 Z5.5 F10.
N951 Y0
N952 Y-2.
N953 X-6.25
N954 Y0
N955 Y2.
N956 G80 Z7.5 M09
N957 G91 G28 Z0
N958 (9/16 EM CRB 4FL 1-1/8 LOC)
N959 T54 M06
N960 S200 M03

N961 ( Rough Mill4 )
N962 G90 G54 G00 X-10.3963 Y.1463
N963 G43 Z7.5 H54 M08
N964 G01 Z6.275 F1.25
N965 Y-.1462 F10.
N966 X-8.1038
N967 Y.1463
N968 X-10.3963
N969 Y.4275
N970 X-10.6775
...

N1209 ( Drill11 )
N1210 G90 G54 G00 X-10.75 Y-5.5
N1211 G43 Z3.1 H55 M08
N1212 G81 G99 R3.1 Z1.75 F10.
N1213 X-9.
N1214 X-7.25
N1215 G80 Z4. M09
N1216 G91 G28 Z0
N1217 (#8 HSS 60Deg Centerdrill)
N1218 T52 M06
N1219 S2800 M03

N1220 ( Center Drill12 )
N1221 G90 G54 G00 X-10.75 Y5.5
N1222 G43 Z4. H52 M08
N1223 G81 G98 R3.1 Z2.3086 F6.
N1224 X-9.
N1225 X-7.25
N1226 G80 Z4. M09
N1227 G91 G28 Z0
N1228 (1 JOBBER DRILL)
N1229 T55 M06
N1230 S1000 M03

N1231 ( Drill12 )
N1232 G90 G54 G00 X-10.75 Y5.5
N1233 G43 Z3.1 H55 M08
N1234 G81 G99 R3.1 Z1.75 F10.
N1235 X-9.
N1236 X-7.25
N1237 G80 Z4. M09
N1238 G91 G28 Z0
N1239 G28 X0 Y0
N1240 M30
```

Figure 6.23 The G-code generated by CAMWorks

You may click the *Post Process* button [Post Process] above the graphics area, and follow the same steps learned in Lesson 2 to convert the toolpath into G-code. Figure 6.23 shows (partial) contents of the NC program (O0001).

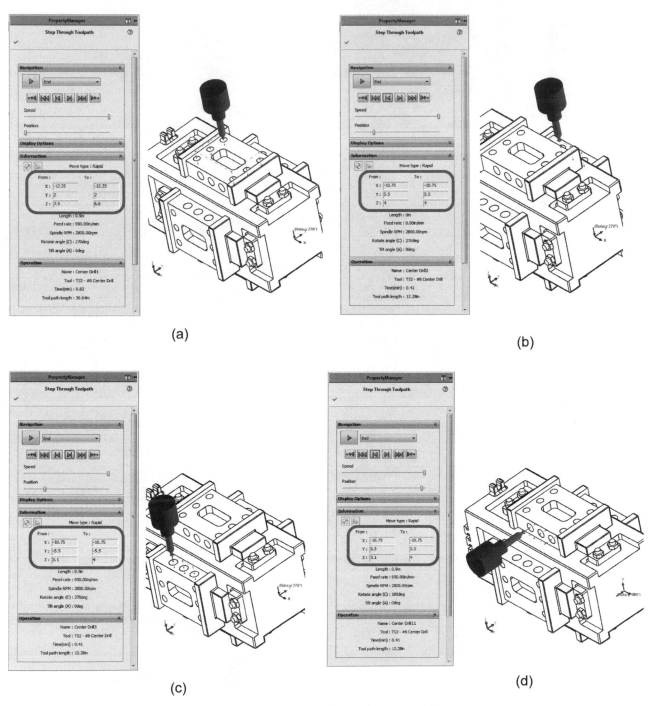

(a)

(b)

(c)

(d)

Figure 6.24 Step through the toolpath: (a) N7 to N14 (Center Drill1), (b) starting blocks N272 (Center Drill2), (c) starting blocks N294 (Center Drill3), and (d) starting blocks N602 (Center Drill6, *Setup2*)

As shown in Figure 6.23, the NC program can be understood in four segments, led by identifier A defining the rotation angle of the 4th axis for the respective four setups. They are NC blocks *N7 A−90*, *N316 A−180*, *N625 A0*, and *N934 A90*, representing *Setup1* to *Setup4*, respectively. These rotation angles are consistent with those shown in Figure 6.18 except that a −90° angle is in place before *Setup1*.

Each segment consists of eight operations. These operations, for example for *Setup1* (N7 −90), start at blocks *N6 (Center Drill1)*, *N20 (Drill1)*, *N34 (Rough Mill1)*, *N201 (Contour Mill1)*, *N271 (Center Drill2)*, *N282 (Drill2)*, *N293 (Center Drill3)*, and *N304 (Drill3)*, for the respective operations.

Blocks N7 and N8 move the drill bit 7.5in. above the center point of the first hole (X = −12.25, Y = 2.0) on the top face of the part—the third hole to the right of the one selected in Figure 6.22(b). Block N9 makes a center drill to the hole, then to the next hole in the middle (X= −12.25, Y = 0), and then to the third hole (X = −12.25, Y = −2.0). This operation can also be seen by using the *Step Through Toolpath* option ⊟, for example, blocks N7 to N14 for *Center Drill1*, as shown in Figure 6.24(a).

Also shown are blocks N272 to N274 for *Center Drill2* [Figure 6.24(b)], blocks N294 to N296 for *Center Drill3* [Figure 6.24(c)], and blocks N602 to N612 shown in Figure 6.24(d) for *Center Drill6* of *Setup2* with 4th axis: A−*180*.

The complete content of the G-code can be seen in the file: *Multiplane Machining with Toolpath.txt*.

We have now completed the lesson. You may save your model for future reference.

6.6   Exercises

Problem 6.1. Create an assembly like that of Figure 6.25 using the parts (*Rotary Table*, *Problem 6.1 Part*, *Clamp 2*, and *Bolt half inch*) and subassembly (*Clamped Part*) in the Problem 6.1 folder downloaded from the publisher's website. Note that the freeform surface of *Problem 6.1 Part* is similar to that of Lesson 4. In addition to the freeform surface, there are holes at the top and side faces of the part. Create a total of four instances as a circular pattern feature.

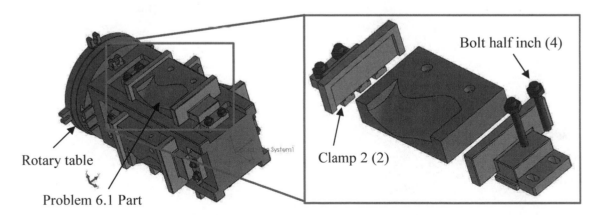

Figure 6.25 Assembly of Problem 6.1

(a)   Generate machining operations for the four instances following steps discussed in this lesson. The operations must cut the freeform surface and holes on the top and side faces. Note that like Lesson 4, you will have to manually create a multisurface machinable feature to machine the freeform surface.

(b) Generate G-codes and verify that the codes are generated correctly by reviewing the contents of the codes, similar to those of Section 6.5.

Problem 6.2. Go back to the example in this lesson and select the *Subprogram* under the *G-code format* option in the *Manage Parts* dialog box (see Figure 6.26). Then, output G-codes. Does the selection of the *Subprogram* under the *G-code format* option output G-codes with subprogram calls in machining the instances? Verify if the G-codes output from CAMWorks are correct.

Figure 6.26 The *Manage Parts* dialog box

Lesson 7: Multiaxis Milling and Machine Simulation

7.1 Overview of the Lesson

In this lesson, we learn to create machining operations that cut a design part with a freeform surface using a 5-axis mill. The freeform surface of the design part is, in particular, a cylindrical surface extruded with a Bézier curve of four control points. In addition to creating a 5-axis milling operation, we will learn to avoid excessive cuts by turning on gouge checking. Moreover, we will learn to bring the toolpath simulation into a CNC machine to carry out the virtual machining simulation in a more realistic setting. As mentioned in Lesson 1, CAMWorks offers Machine Simulation capability that employs virtual CNC machines to support users in carrying out machining simulation. In this lesson, we will use *Mill_Tutorial*, one of the two generic virtual machines that come with CAMWorks, to learn Machine Simulation capability. In addition to simulating machining operations, Machine Simulation supports tool collision detection in a more realistic setting. More importantly, this capability allows you to add your own virtual CNC machine, a virtual replica of the physical machine available at your machine shop, into CAMWorks to carry out virtual machining simulation. We will discuss the steps of adding a 5-axis HAAS mill to CAMWorks in Lesson 10.

We create three operations to cut the part, similar to those of Lesson 4, including a volume milling, a local milling, and a surface milling operation. In Lesson 4 we assumed a 3-axis mill. In this lesson, we use 5-axis mill for surface milling, as shown in Figure 7.1(a). Thereafter, we change the surface milling sequence using a larger flat-end cutter to discuss the capability in CAMWorks that adjusts the toolpath to avoid gouging. The toolpath will also be simulated using a legacy machine, *Mill_Tutorial*, in Machine Simulation, as shown in Figure 7.1(b). This legacy machine is similar to the work chamber of a typical 5-axis mill with a tilt table and a rotary table that provide rotations in X and Z-axes, respectively; for example, a HAAS VF-5 CNC mill shown in Figure 7.1(c).

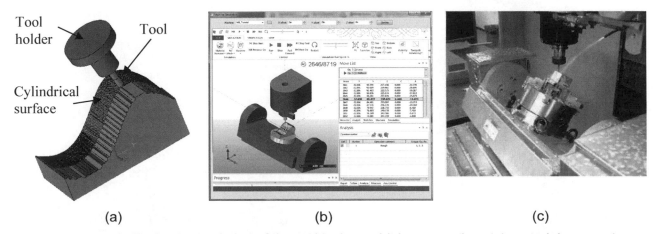

| (a) | (b) | (c) |

Figure 7.1 Toolpath simulation of the multiaxis machining operation, (a) material removal simulation, (b) virtual machine simulation, and (c) HAAS VF-5 CNC mill

At the end of this lesson, we re-visit the surface milling operation of the example in Lesson 4 and explore the feasibility of using a 5-axis surface milling operation to improve the accuracy and quality of the machined freeform surface.

7.2   The Multiaxis Machining Example

The design part (filename: *Cylindrical Surface.SLDPRT*) consists of a solid extrusion feature, see Figure 7.2(a), with a sketch of three straight lines and a Bézier curve of four control points, as shown in Figure 7.2(b). These four control points labeled $\mathbf{P}_0$, $\mathbf{P}_1$, $\mathbf{P}_2$, and $\mathbf{P}_3$ are dimensioned so that their X-Y coordinates are defined on the sketch plane as (0, 3), (4, 9), (5, −2) and (10, 2), respectively. The size of the bounding box of the solid feature is 10in.×4.833in.×5in. The stock employed for this lesson is a rectangular block made of Steel 1005, of 10in.×5.083in.×5in., and a 0.25in. extension in the +Y direction from its bounding box. The extrusion of the Bézier curve generates a cylindrical surface, in which the quality of the machined surface would be desirable by implementing the three-operation scenario similar to that of Lesson 4 but with a 5-axis mill for the surface milling operation.

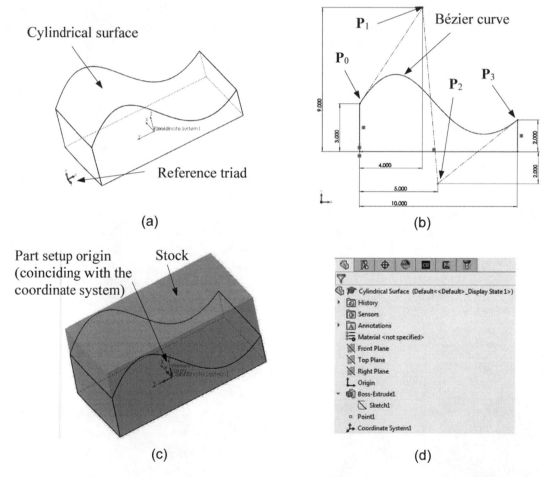

(a)   (b)

(c)   (d)

Figure 7.2 The design part, *Cylindrical Surface.SLDPRT*, (a) the solid model and the cylindrical surface, (b) the Bézier curve in sketch, (c) stock and part setup origin, and (d) the *FeatureManager* design tree

A coordinate system, *Coordinate System1*, is defined with its origin at the center of the bottom face of the solid model (*Point1*). Like previous lessons, *Coordinate System1* is chosen as the fixture coordinate system, again defining the "home point" or main zero position on the machine. Unlike previous lessons, the X-, Y-, and Z-axes of *Coordinate System1* point in different directions as those of the reference triad shown at the lower left corner of the graphics area. X-, Y-, and Z-axes of the fixture coordinate system must be defined consistently with those of the virtual machine in Machine Simulation, in which X- and Z-axes align with those of the tilt table and rotary table, respectively. This will become clear in Section 7.7 when we discuss Machine Simulation. The origin of the coordinate system will be chosen as the part setup origin; see Figure 7.2(c). Again, a part setup origin defines the G-code program zero location.

The unit system chosen is IPS (inch, pound, second). When you open the solid model *Cylindrical Surface.SLDPRT*, you should see the solid feature (*Boss-Extrude1*), a point, and a coordinate system listed in the *FeatureManager* design tree like that of Figure 7.2(d).

There are three operations to be created for this example. The first operation is a volume milling (*Area Clearance*), which is a rough cut using a 2in. flat-end mill. The toolpath of the volume milling operation can be seen in Figure 7.3(a). The second operation is a local milling that continues removing material remaining from the volume milling using a smaller ball-nose cutter of diameter 0.5in.; see toolpath in Figure 7.3(b). The third operation is a multiaxis surface milling serving as a finish cut that intends to improve the quality and accuracy of the machined part to meet the quality and accuracy requirements; see toolpath in Figure 7.3(c).

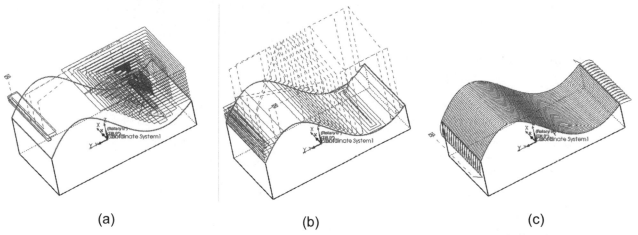

(a) (b) (c)

Figure 7.3 Toolpath of the three NC operations: (a) volume milling, (b) local milling, and (c) multiaxis surface milling

You may open the example file with toolpath created (filename: *Cylindrical Surface with toolpath.SLDPRT*) to preview the toolpaths created for this example. When you open the file, you should see the four operations listed under CAMWorks operation tree tab ![tab icon], as shown in Figure 7.4. Note that the operation *Multiaxis Surface Milling Gouge Checking* is suppressed. You may un-suppress it to review a toolpath that avoids gouging. You may simulate individual operations by right clicking the entity and choosing *Simulate Toolpath*, or simulate the combined operations by clicking the *Simulate Toolpath* button ![Simulate Toolpath] above the graphics area.

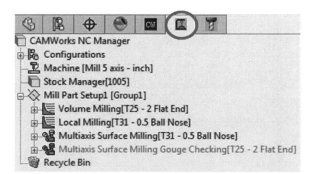

Figure 7.4 The NC operations listed under the CAMWorks operation tree tab

You may also right click *Mill Part Setup1* and choose *Machine Simualtion > Legacy* to bring up the *Machine Simulation* window like that of Figure 7.1(b).

7.3   Using CAMWorks

Open SOLIDWORKS Part

Open the part file (filename: *Cylindrical Surface.SLDPRT*) downloaded from the publisher's website. This solid model, as shown in Figure 7.2(d), consists of one extrude solid feature, a point, and a coordinate system. As soon as you open the model, you may want to check the unit system chosen and make sure the IPS system is selected. You may also increase the decimals from the default 2 to 4 digits similar to that of the previous lesson. Since steps in creating the first two operations are similar to those of Lesson 4, some screen captures of dialog boxes will be skipped to minimize repetitions.

Select NC Machine

Click the CAMWorks feature tree tab ⬛ and right click *Mill-in* to select *Edit Definition*. In the *Machine* dialog box, we select *Mill 5 axis-inch* under *Machine* tab—Figure 7.5(a)—and click *Select*, choose *Crib 1* under *Available tool cribs* of the *Tool Crib* tab, select *M5AXIS-TUTORIAL* under the *Post Processor* tab, Figure 7.5(b), and select *Coordinate System1* under *Fixture Coordinate System* of the *Setup* tab.

Next, we choose Z-axis as the rotary axis, and X-axis as the tilt axis.

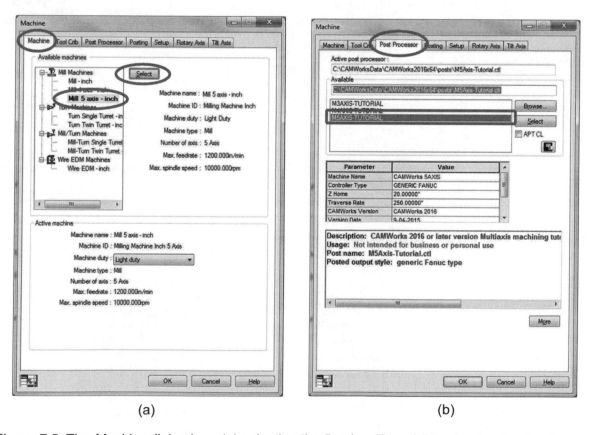

(a)   (b)

Figure 7.5 The *Machine* dialog box, (a) selecting the 5-axis mill, and (b) selecting the 5-axis post-processor

In the *Machine* dialog box, we choose the *Rotary Axis* tab, and select *Z axis* as the rotary axis; see Figure 7.6(a). Then, we choose the *Tilt Axis* tab, and select *X axis* as the tilt axis, and *XY plane* for the *0 degree position*; see Figure 7.6(a). We click *OK* to accept the definition.

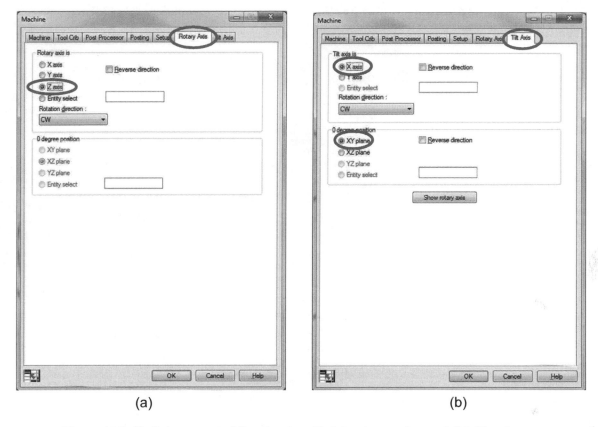

(a)   (b)

Figure 7.6 Defining axes of the 5-axis mill, (a) rotary axis, and (b) tilt axis

Create Stock

From CAMWorks feature tree ![CW], right click *Stock Manager* and choose *Edit Definition*. In the *Stock Manager* dialog box, we increase the Y+ field from 0 to 0.25in. and keep the default material (*Steel 1005*). A rectangular stock of 10in.×5.083in.×5in. should appear in the graphics area similar to that of Figure 7.2(c).

Next, we select the cylindrical surface to create a machinable feature since the cylindrical surface is not one of the standard 2.5 axis features recognizable by AFR.

Create a Machinable Feature

Click the CAMWorks feature tree tab ![CW], right click *Stock Manager*, and choose *New Mill Part Setup*, as shown in Figure 7.7.

The *Mill Setup* dialog box appears (Figure 7.8), select the *Top Plane* and click the *Reverse Selected Entity* button ![icon] under *Entity* to reverse the direction. Make sure that the arrow of the tool axis symbol ![icon] points in a downward direction, as shown in Figure 7.8. Click the checkmark ✔ to accept the definition. A *Mill Part Setup1* is now listed in the feature tree.

Now we define a machinable feature. From CAMWorks feature tree ▣, right click *Mill Part Setup1* and choose *New Multi Surface Feature* (Figure 7.9). The *New Multi Surface Feature* dialog box appears (Figure 7.10).

Pick the cylindrical surface of the part in the graphics area; the surface picked is now listed under *Selected Faces*. Leave the default *Area Clearance, Pattern Project* for *Strategy*, and click the checkmark ✔ to accept the machinable feature.

In the CAMWorks feature tree ▣, a *Multi Surface Feature1* is added in magenta color under *Mill Part Setup1*.

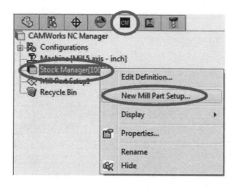

Figure 7.7 Selecting *New Mill Part Setup*

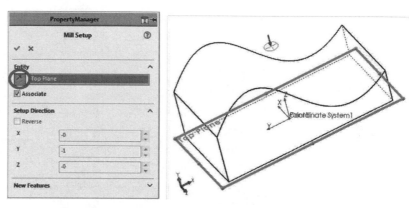

Figure 7.8 Picking the top face for mill setup

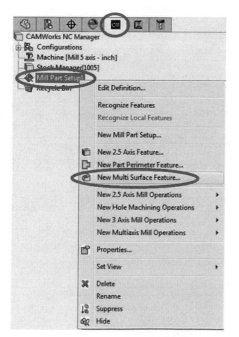

Figure 7.9 Choosing *New Multi Surface Feature*

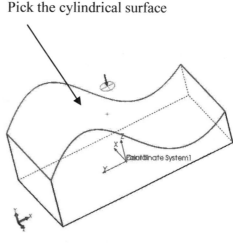

Pick the cylindrical surface

Figure 7.10 Picking the cylindrical surface for machinable feature

Generate Operation Plan and Toolpath

Click the *Generate Operation Plan* button above the graphics area. Two operations, *Area Clearance1* and *Pattern Project1*, are generated. They are listed in CAMWorks operation tree. Again they are shown in magenta color. Change the name of the operation from *Area Clearance* to *Volume Milling*.

The part setup origin and axis coincide with the fixture coordinate system by default. We will stay with the default selection. You may review them by right clicking *Mill Part Setup1* and choosing *Edit Definition*.

In the *Part Setup Parameters* dialog box, see Figure 7.11(a); the *Fixture coordinate system* under the *Origin* tab has been chosen. Click the *Axis* tab, as shown in Figure 7.11(b); again the *Fixture coordinate system* has been chosen. The default selections set the part setup origin and axes to coincide with the coordinate system, *Coordinate System1*, as shown in Figure 7.2(c). Click *Cancel* to close the dialog box.

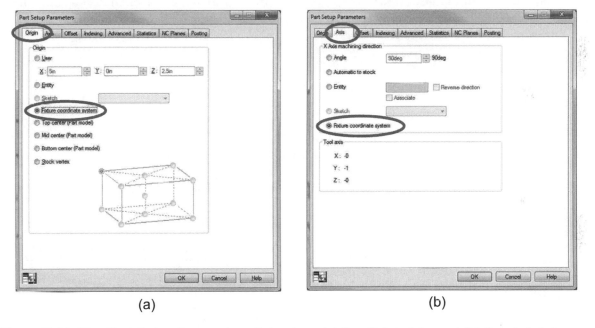

(a)   (b)

Figure 7.11 The *Part Setup Parameters* dialog box, (a) the *Origin* tab, and (b) the *Axis* tab

Click the *Generate Toolpath* button above the graphics area to create the toolpath. The two operations are turned into black color right after toolpaths are generated.

We change the tool of the *Volume Milling* operation to a 2in. flat-end cutter. This can be done by right clicking *Volume Milling* and choosing *Edit Defintion*.

In the *Operation Parameters* dialog box, as shown in Figure 7.12(a), choose *Tool Crib* tab and choose *Filter* to display flat-end tools only.

Choose the 2in. flat-end cutter—*ID:721* in Figure 7.12(a)—and click *Select*. Click *Yes* to the question: *Do you want to replace the corresponding holder also?*

We have now replaced the cutter with a 2in. flat-end cutter. The selected cutter should appear under the *Tool* tab; see Figure 7.12(b).

Choose the *Area Clearance* tab, and note that the *Allowance* and *Z allowance* are both set to 0.01in. under *Surface finish*, as shown in Figure 7.12(c). The allowances leave 0.01in. thickness meterial on the cylindrical surface. This thin layer material minimizes (or eliminates) tool marks on the finished surface that improves the quality of the machined part.

We use default machining parameters determined by the TechDB™ for this exercise.

Click *OK* to accept the changes and click *Yes* to the warning message: *Operation parameters have changed, toolpaths need to be recalculated. Regnenerate toolpaths now?*

The toolpath will be generated like that shown in Figure 7.3(a).

Right click *Volume Milling* and choose *Simulate Toolpath* to show the material removal simulation of the operation, like that of Figure 7.13(a). As shown in Figure 7.13(a), there is a significant amount of material uncut. Next, we add a local milling operation to further remove the material remaining on the cylindrical surface. Also, we delete *Pattern Project1* by right clicking it and choosing *Delete*. We will create a multiaxis surface milling operation later.

7.4 Adding a Local Milling Operation

Similar to Lesson 4, we add a local milling operation by creating another area clearance operation that continues machining the material remaining from the *Volume Milling* operation with a ball-nose cutter of 0.5in. diameter, and 0.2in. for both stepover and depth of cut.

Right click *Mill Part Setup1* and choose *New 3 Axis Mill Operations > Area Clearance*.

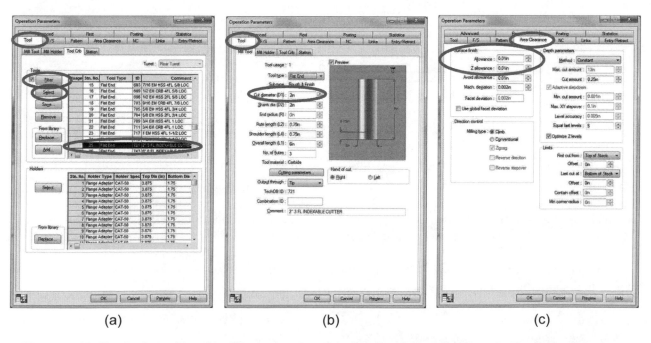

| (a) | (b) | (c) |

Figure 7.12 Replacing with a 2in. flat-end cutter, (a) showing flat-end cutters in the tool crib using the filter option, (b) 2in. flat-end cutter selected, and (c) *Allowance* under the *Area Clearance* tab

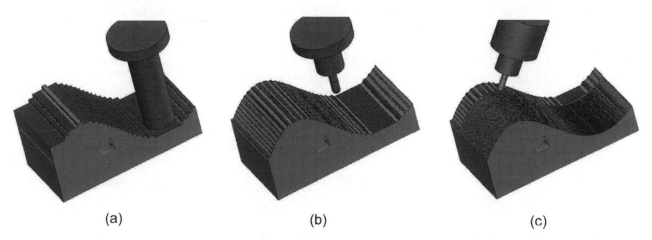

(a)   (b)   (c)

Figure 7.13 Material removal simulation of the three NC operations, (a) volume milling, (b) local milling, and (c) multiaxis surface milling

In the *New Operation: Area Clearance* dialog box (Figure 7.14), click the *Select Multi Surface Feature* button (circled in Figure 7.14). Select *Multi Surface Feature1* in the *Features for Area Clearance Operation* dialog box (Figure 7.15). The selected entity is listed under *Features* of the *New Operation: Area Clearance* dialog box (Figure 7.14). Click the checkmark ✓ to accept the new operation.

In the *Operation Parameters* dialog box, choose *Tool Crib* tab, and display only ball-nose cutters using the filter option. Choose the 0.5in. cutter (*ID: 786, 1/2 HSS 2FL*) as seen in Figure 7.16(a), and click *Select*. Click *Yes* to replace the tool holder.

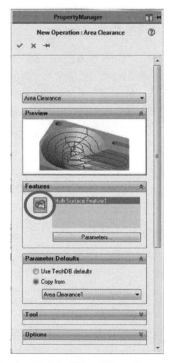

Figure 7.14 The *New Operation:*
Area Clearance dialog box

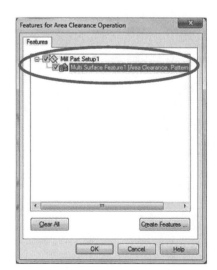

Figure 7.15 Selecting *Multi Surface*
Feature1

Choose the *Pattern* tab of the *Operation Parameters* dialog box, select *Pocket Out* for *Pattern*, enter 40% and 20% for *Max stepover%* and *Min stepover%*, as shown in Figure 7.16(b).

Choose the *Area Clearance* tab of the *Operation Parameters* dialog box; see Figure 7.16(c). In the *Depth* parameters group, enter *0.2in* for *Cut amount*, as shown in Figure 7.16(c).

Choose the *Rest* tab of the *Operation Parameters* dialog box, as shown in Figure 7.17(a). Choose *From WIP* for *Method*, and click the selection button ▣ to select *Rough Mill* for *Compute WIP from operations*. This is how we link the local milling to a previous operation.

Rename *Area Clearance2* as *Local Milling*. Right click *Local Milling* and choose *Generate Toolpath* to generate a toolpath like that shown in Figure 7.3(b).

Simulate the toolpaths by clicking the *Simulate Toolpath* button ▣ above the graphics area. Click the *Run* button ▶ to simulate the toolpath. The material removal simulation of the two operations appear in the graphics area, similar to that of Figure 7.13(b).

7.5 Adding a Multiaxis Surface Milling Operation

We will add a multiaxis surface milling operation to polish the cylindrical surface of the machined part using the same 0.5in. ball-nose cutter as that of the local milling operation.

Right click *Mill Part Setup1* and choose *New Multi Axis Operations > Multi Axis Mill*.

In the *New Operation: Multiaxis Mill* dialog box (Figure 7.18), click the *Select Multi Surface Feature* button (circled in Figure 7.18). Select *Multi Surface Feature1* in the *Features for Multiaxis Mill Operation* dialog box (see Figure 7.15). The selected entity is listed under *Features* of the *New Operation: Multiaxis Mill* dialog box (Figure 7.18). Click the checkmark ✔ to accept the new operation.

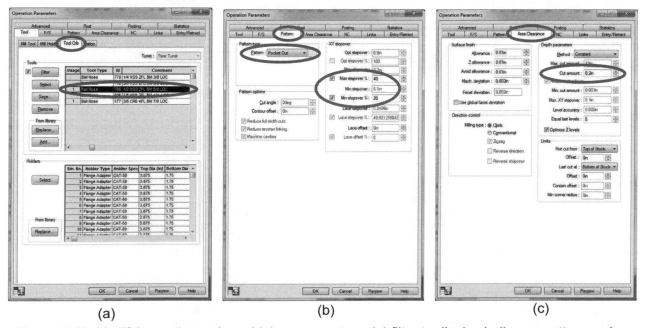

(a)   (b)   (c)

Figure 7.16 Modifying cutter and machining parameters, (a) filter to display ball nose cutters under the *Tool Crib* tab, (b) edit parameters and options under the *Pattern* tab, and (c) change cut amount under the *Area Clearance* tab

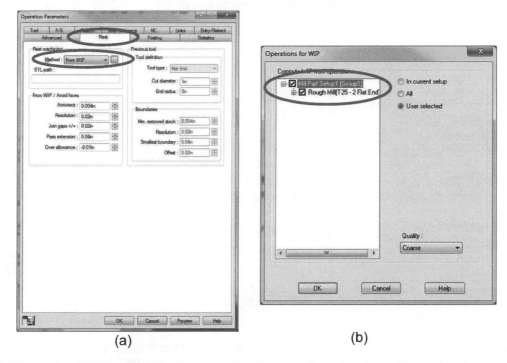

(a)   (b)

Figure 7.17 Linking the local milling to the previous operation, (a) choosing method under the *Rest* tab, and (b) picking *Rough Mill* in the *Operations for WIP* dialog box

In the *Operation Parameters* dialog box, choose the 0.5in. cutter (*ID: 786, 1/2 HSS 2FL*) following the same steps as before.

Click the *Pattern* tab, as shown in Figure 7.19(a), to review the pattern type and stepover. Change the pattern to *Flowline Between Curves*.

Click *Upper*; the *Curve Wizard*, as shown in Figure 7.19(b), appears. Pick the front Bézier curve of the cylindrical surface; see Figure 7.19(c) in the graphics area. Notice that the point at the left end of the curve is highlighted, indicating the start point of the curve.

Click *Lower*, and pick the Bézier curve at the rear side of the cylindrical surface in the graphics area. Notice that the point at the left end of the curve is highlighted, indicating the start point of the curve. The start points of the curves are consistent leading to a valid toolpath. Choose S*tart and End At Exact Surface Edge* for *Method* under *Limits*. Change the *Max. scallop* to 0.01 in.

Figure 7.18 The *New Operation: Multi Axis Mill* dialog box

Click the *Axis Control* tab (Figure 7.20). Make sure *5 Axis* is chosen, and select *Normal to Surface* for *Tool axis*.

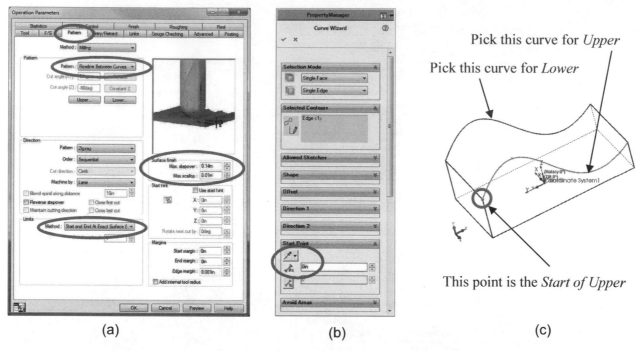

(a)   (b)   (c)

Figure 7.19 Defining a *Flowline* type operation, (a) choosing *Flowline* for *Pattern* type, (b) the curve wizard showing single edge, curve picked, and options of selecting start point, and (c) picking curves

Click the *Preview* button to show the toolpath like that of Figure 7.3(c). We accept the toolpath. Click the close button ⊠ at the top right corner of the *Operation Parameters* dialog box, and click *OK* to accept the changes and regenerate the toolpath.

Simulate the toolpaths by clicking the *Simulate Toolpath* button ▣ above the graphics area. Click the *Run* button ▶ to simulate the toolpath. The material removal simulation of the operations appear in the graphics area, similar to that of Figure 7.13(c). We rename the operation as *Multiaxis Surface Milling*.

7.6   Tool Gouging

Next, we replace the tool with a 2in. flat-end cutter for the surface milling operation, which is certainly inadequate. We make this change to illustrate the gouge checking capability of CAMWorks.

Right click *Multiaxis Surface Milling* and choose *Edit Definition*. Choose the 2in. flat-end mill (Tool ID: 721) under *Mill Tool* tab. Change the *Max stepover* to 1in. under the *Pattern* tab. Regenerate the toolpath.

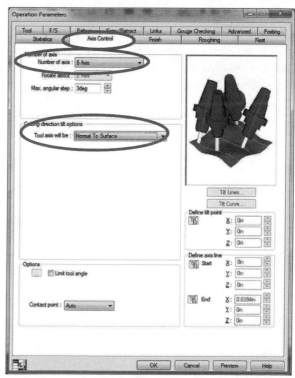

Figure 7.20 Defining tool axis control

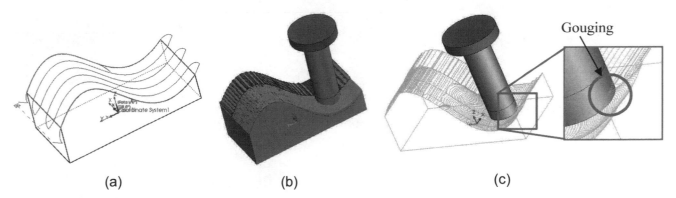

(a)   (b)   (c)

Figure 7.21 Illustration of tool gouging, (a) toolpath of the multiaxis surface milling operation using a 2in. flat-end cutter, (b) material removal simulation, and (c) tool gouging area and a zoom-in view

The toolpath is shown in Figure 7.21(a), and the material removal simulation of the combined three operations shown in Figure 7.21(b) seems to be fine. However, after taking a closer look; for example, rotate the view to that of Figure 7.21(c), choose *Stock Wireframe Display* and *Target Part No Display*, from the *Toolpath Simulation* tool box; the cutter gouges into the material at the concave area of large curvature, making an excessive over cut.

Gouging is highly undesirable. How do we eliminate or avoid gouging in toolpath generation?

Right click *Multiaxis Surface Milling* and choose *Edit Definition*. Choose the *Gouge Checking* tab (see Figure 7.22), select *Group1*, apply gouge checking to *Holder*, *Shank*, *Non-cutting portion*, and *Flute*. Choose *Move Tool Away* for *Strategy* and *Along Surface Normal* for *Retract tool*. Click *OK*, and regenerate the toolpath.

The regenerated toolpath is shown in Figure 7.23(a), showing that the tool is lifted up from the cylindrical surface at the concave area, where gouges occurred previously. The material removal simulation shown in Figure 7.23(b) indicates that material at the concave area is mostly uncut due to the lift of tool in the area to avoid gouging. A closer look at material removal simulation verifies that the tool is lifted up in the area to avoid gouging; see Figure 7.23(c).

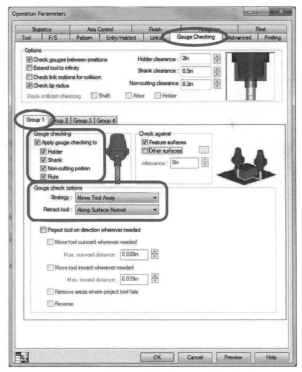

Figure 7.22 Defining gouge checking in the *Operation Parameters* dialog box

7.7   Machine Simulation

Machine Simulation in CAMWorks offers a more realistic setup in simulating machining operations. For example, a sample machine simulator called *Mill_Tutorial* (among others provided by CAMWorks such as *MillTurn_Tutorial*), provides a setup of tools with a tool holder and stock mounted on a tilted rotary table, as shown in Figure 7.1(b).

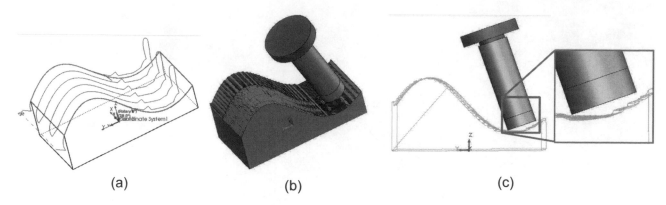

(a)   (b)   (c)

Figure 7.23 Toolpath avoiding gouging, (a) toolpath of the modified multiaxis surface milling operation, (b) material removal simulation, and (c) tool lift up from the cylindrical surface

Machine Simulation also detects tool collisions when they happen. Note that you may add your own virtual CNC machines, duplicates of the physical machines at your machine shop, to CAMWorks that allows you to carry out more realistic machining simulation. You may also add associated post processors to CAMWorks that support generating G-codes compatible with the respective CNC machines. More about adding virtual machines to CAMWorks will be discussed in Lesson 10.

To bring up the Machine Simulation window, you may right click *Mill Part Setup* under CAMWorks operation tree ![icon], and choose *Machine Simulation* and then *Legacy*.

The basic layout of the Machine Simulation window is shown in Figure 7.24. In the graphics area, the cutter, tool holder, tilt table, rotary table, and the stock together with a machine coordinate system XYZ are displayed. To the right, the three machining operations and corresponding G-codes are listed in the upper and lower areas of *Move List*, respectively. On top, the default simulator, *Mill_Tutorial*, is selected, with offsets in X-, Y-, and Z-directions all set to zero. Below are buttons that control the machining simulation run. Under these buttons are four tabs, *FILE*, *SIMULATION*, *VERIFICATION*, and *VIEW*, supporting respective uses of the machine simulation capability. *SIMULATION* tab is chosen as default. Under the *SIMULATION* tab, there are *Simulation* options that offer setups to display the desired simulation, such as material removal. Next to the *Simulation* options are buttons that provide options for *Control*, *Simualtion Run Speed*, and *Views*.

Before we start, click the *Select* button ![icon] next to the *Machine* on top of the *Machine Simulation* window (Figure 7.24) to bring up the *Select Point* dialog box (Figure 7.25). We are about to select the origin of the work coordiante system on the part that determines its position and orientation with respect to the top face of the rotary table. Pick the coordinate system, *Coordinate System1*, in the graphics window (see Figure 7.26), to choose the origin of *Coordinate System1* as the origin of the work coordinate system. And then, click the *Update* button on top. The origin of *Coordinate System1* (at the center of the bottom face of the stock) coincides with the center point at the top face of the rotary table (see Figure 7.24). Also, the offsets in X-, Y-, and Z-directions are all zero.

To verify if a correct coordinate system has been chosen for the Machine Simulation, you may choose *TableTable* as the machine by pulling down the *Machine* selection and choosing it (see Figure 7.27). Click the *Update* button. In the graphics area, a work coordinate system appears at the center of the bottom face of the stock with its X-, Y-, and Z-axes parallel to those of the virtual machine shown at the lower left corner of Figure 7.28.

Now, we choose *Mill_Tutorial* as the machine and click the *Update* button, and run the simulation.

Select button

Tool holder

Cutter

Tilt table

Rotary table

Stock

Update button

NC sequences

NC codes

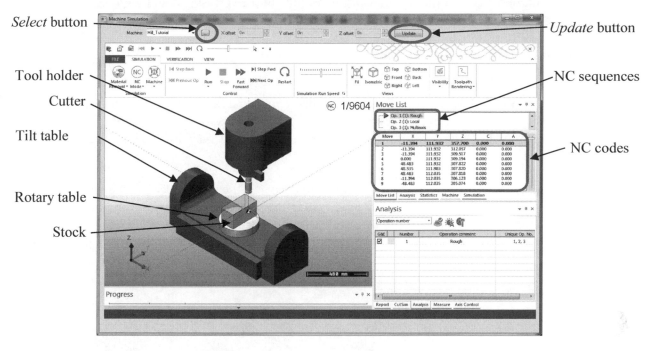

Figure 7.24 The *Machine Simulation* window

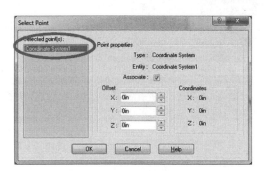

Figure 7.25 The *Select Point* dialog box

Pick *Coordinate System1*

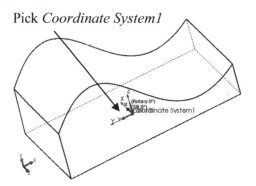

Figure 7.26 Picking the coordinate system, *Coordinate System1*

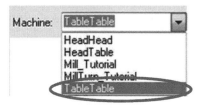

Figure 7.27 Choosing *TableTable* for machine

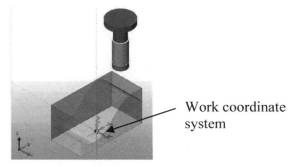

Work coordinate system

Figure 7.28 The work coordinate system displayed

Click the *Run* button ▶ to start the simulation. The G-code of the corresponding tool motion is displayed in the lower portion of the *Move List* area.

The simulation runs to the end for all three operations without any issue. There is no collision encountered.

Close the *Machine Simulation* window. Suppress the first two operations by right clicking *Volume Milling* (and *Local Milling*) and choosing *Suppress* under the CAMWorks operation tree 🖫. Rerun the machine simulation. The simulation starts from the third operation, *Multiaxis Surface Milling*, directly since the first two operations were suppressed.

Without the volume milling and local milling, a collision between the tool holder and the stock is detected immediately after the simulation starts. Click *Yes to All* in the warning box, as shown in Figure 7.29(a), to allow the simulation to continue. The collison area is highlighted in red—Figure 7.29(b)—and a symbol ✖ appears in the problematic NC blocks in the G-code where collisions are detected; see Figure 7.29(c).

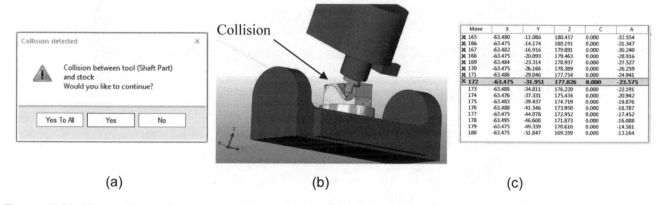

(a)   (b)   (c)

Figure 7.29 The collision detection in *Machine Simulation*, (a) the *Collision detected* warning box, (b) collision area highlighted, and (c) the symbol ✖ appearing in front of problematic NC blocks

Close the *Machine Simulation* window. Unsuppress the first two operations by right clicking *Volume Milling* (and *Local Milling*) and choosing *Unsuppress* under the CAMWorks operation tree 🖫.

We have completed the exercise. You may save the model for future reference.

Next, we revisit the freeform surface exmaple of Lesson 4. We create a multiaxis surface milling operation using a 5-axis mill in place of the *Pattern Project* operation in hopes of eliminating or reducing the noticable material remaining at numerous areas; for example, near the middle of the front edge, as indicated in Figure 4.36 of Lesson 4.

7.8   Revisiting the Freeform Surface Example

We now open the example file saved from Lesson 4, in which three operations were generated, *Volume Milling*, *Local Milling*, and *Surface Milling,* using a 3-axis mill.

Under CAMWorks operation tree 🖫, right click *Mill Part Setup1*, and choose *New Multiaxis Mill Operation > Multiaxis Mill.*

We follow the same steps outlined in Section 7.5 to create a multiaxis surface mill operation. That is, we pick the freeform surface as the machinable feature, and choose a 0.5in. ball-nose cutter (T54).

Under the *Pattern* tab of the *Operation Parameters* dialog box (see Figure 7.30), we choose *Slice* for *Pattern*, choose *Start and End At Exact Surface Edges* for *Method*, and enter *0.001* for *Max scallop*.

We define tool axis control using the exact same options as those of Figure 7.20; i.e., 5 axis and normal to surface.

The resulting toolpath of the multiaxis surface milling is shown in Figure 7.31(a). Because of the option we chose, *Start and End At Exact Surface Edges* for *Method*, and the fact that a 5-axis mill is employed, the toolpaths show that two passes of the toolpath stay right on the two respective boundary edges of the freeform surface. As a result, the noticable material near the boundary edges seen in Lesson 4 will be removed.

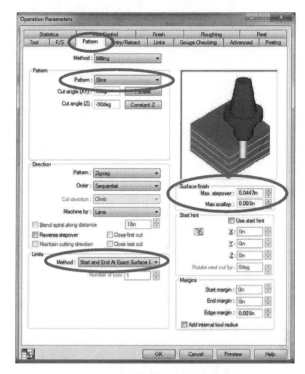

Figure 7.30 Defining a 5-axis surface mill operation under the *Pattern* tab

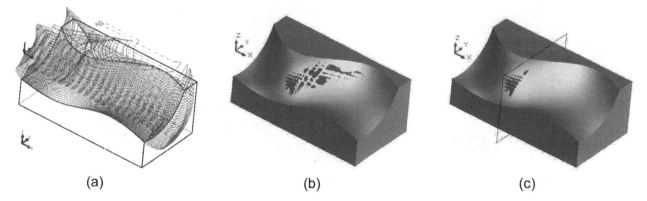

(a)   (b)   (c)

Figure 7.31 Toolpath and material removal simulation of the freeform surface example, (a) toolpath of the multiaxis surface milling operation, (b) material removal simulation of the combined machining operations, and (c) section view at offset 3.2in. in YZ plane

The machining simulation of the combined three operations is shown in Figure 7.31(b) and Figure 7.31(c) (section view). The noticable amount of material uncut near the front edge as seen in Lesson 4 is completely eliminated, as predicted. Same is true near the top edge. Note that the areas of purple color appearing on the freeform surface is simply due to visualization. Rotating the view slighlty leads to a machined surface without purple areas (see Figure 7.32).

We have completed revisiting this example. You may save your model for future reference.

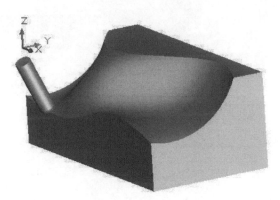

Figure 7.32 The machined surface at the end of the material removal simulation of combined toolpaths in a slightly rotated view

7.9   Exercises

Problem 7.1. In this exercise, we rework the freeform surface example of Problem 4.1.

(a)   Replace the *Pattern Project* operation by creating a multiaxis surface milling operation using a 5-axis mill. Would a 5-axis mill offer a better toolpath than that of *Pattern Project*, especially near the front edge where notable material remained uncut?

(b)   Adjust the allowance parameters of the *Volume Milling* and *Local Milling* operations to undersand the effect of the parameters to the quality of the machined surface. For example, set allowances to zero for both operations. Generate toolpath and simulate machining for all three operations. Can you identify the undesired tool marks on the freeform surface?

(c)   Simulate the machining operations using the *Machine Simulation* capability, similar to that of Figure 7.33.

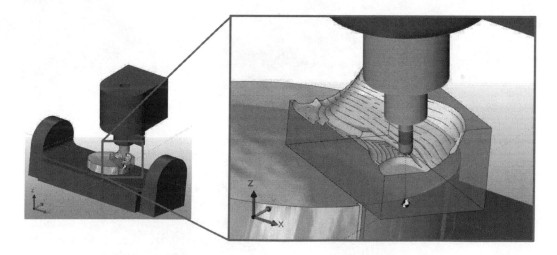

Figure 7.33 Machine simulation of Problem 7.1

Lesson 8: Turning a Stepped Bar

8.1   Overview of the Lesson

We discuss turning operations in Lessons 8 and 9. In Lesson 8, we use a simple stepped bar example to learn basic capabilities in generating turning operations and understanding G-codes post-processed by CAMWorks. In Lesson 9, we machine a similar example with more turning features to gain a broader understanding of turning capabilities offered by CAMWorks.

This current lesson provides you with a quick run-through in creating turning operations using CAMWorks. You will learn a complete procedure in using CAMWorks to create turning operations from the beginning all the way to the post process that generates the G-code. We use a lathe of single turret to machine the simple stepped bar example shown in Figure 8.1(a) from a bar stock clamped into a three-jar chuck—see Figure 8.1(b)—by creating an outer profile turning operation (called outer diameter or OD turning in CAMWorks) and a cut off operation (removing the part from the stock). The machined part at the end of the material removal simulation is shown in Figure 8.1(c). This lesson is intentionally made simple. We stay with default options and parameters for most of the lesson.

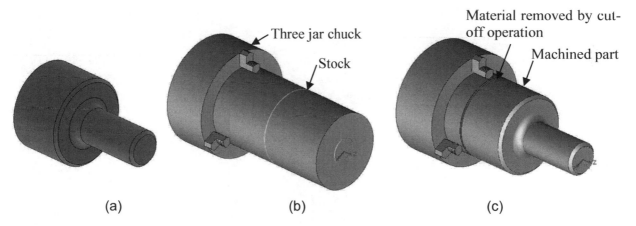

(a)   (b)   (c)

Figure 8.1 The stepped bar example, (a) the design model, (b) bar stock clamped to a chuck, and (c) the material removal simulation

Similar to the milling lessons, we will follow the general steps shown in Figure 1.1 of Lesson 1 to turn the stepped bar. We will start by (1) opening the stepped bar design model; (2) defining machine setup, in which we choose a single turret lathe, select a tool crib, pick a post processor, and choose a fixture coordinate system; (3) creating a stock, in this case, a cylinder enclosing the design model; (4) defining machinable features using interactive feature recognition (IFR) and later automatic feature recognition (AFR) in Lesson 9; (5) generating operation plans and toolpaths; and (6) checking results and reviewing material removal simulation. In addition, we will go over a post process to create G-codes of selected operations, including turn finish and cut off. We will take a closer look at the G-codes generated to gain a better understanding of the capability.

After completing this lesson, you should be able to carry out machining simulation for similar problems following the same procedures and be ready to move onto Lesson 9.

8.2   The Stepped Bar Example

The stepped bar shown in Figure 8.2(a) has a bounding cylinder of size ϕ4.25in.×6.5in. The unit system chosen is IPS. There is one revolve feature created by revolving the sketch shown in Figure 8.2(b). In addition, there is a chamfer of 0.15in. (defined at two edges shown in Figure 8.2(a)) and a fillet (0.375in. in radius) features, plus a point and a coordinate system. These features are listed in the feature tree shown in Figure 8.3. The coordinate system (*Coordinate System1*) and the point (*Point1*) are chosen as the fixture coordinate system and the part setup origin, respectively, for turning operations. When you open the solid model *Stepped bar.SLDPRT*, you should see the solid and the reference features listed in the feature tree (Figure 8.3).

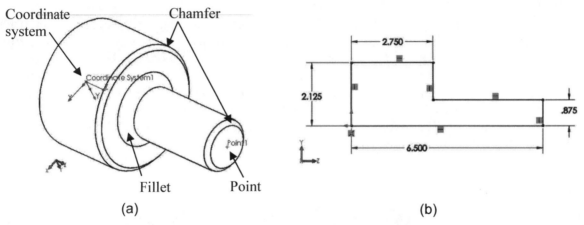

(a)   (b)

Figure 8.2 The stepped bar design model, (a) solid and reference features, and (b) sketch of the revolved solid feature

A bar stock of size ϕ4.25in.×7.75in., made of low carbon alloy steel (1005), as shown in Figure 8.4(a), is chosen for the machining operations. The front end of the stock aligns with the front face of the part. The rear end of the stock is clamped in the three-jar chuck with an extra length of 1.25in. to be cut off at the end of the turning operation. Since the front end faces align, there is no need to define a face turning operation.

Note that a part setup origin is defined at the center of the front end face of the stock (coinciding with *Point1* defined in part), which defines the G-code program zero location.

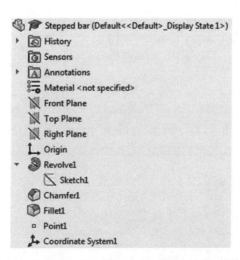

Figure 8.3 Entities listed in the feature tree

We will define two machinable features, an OD (outer diameter) feature that includes the outside shape of the part from the front end face to the cut off face, and a cut off feature to trim off excess stock from the back of the part. In general, an OD machinable feature excludes the shape of any groove features (which are not present in this simple example).

We first define these machinable features manually, using the interactive feature recognition (IFR) capability. In the next lesson, we will use the automatic feature recognition (AFR) capability to recognize machinable features from the part solid model. In this lesson, we follow the recommendations of the technology database (TechDB™) for choosing machining options, cutters, and cutting parameters.

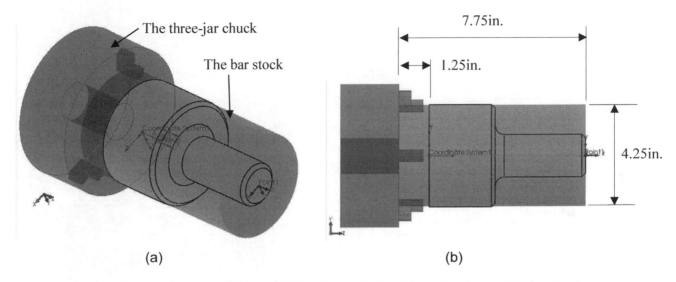

(a)   (b)

Figure 8.4   A bar stock of ϕ4.25in.×7.75in., (a) stock fixed in a chuck, and (b) stock size

The toolpaths of the turning operations for machining the OD feature, consisting of *Turn Rough* and *Turn Finish*, are shown in Figure 8.5(a) and Figure 8.5(b), respectively. The turn rough toolpath moves the cutter along the outer profile of the part in numerous passes. This is because the cut amount is chosen as 0.1in. and the overall depth to cut is 1.25in. The turn finish toolpath moves the cutter along the part profile in one pass, similar to the contour mill operation in milling. The cut off toolpath simply moves the cutter along the negative X direction to separate the part from the stock, as shown in Figure 8.5(c).

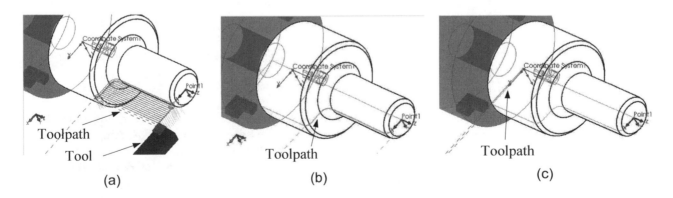

(a)   (b)   (c)

Figure 8.5   The toolpaths of the turning operations, (a) *Turn Rough*, (b) *Turn Finish*, and (c) *Cut Off*

8.3   Using CAMWorks

Open SOLIDWORKS Part

Open the stepped bar design model (filename: *Stepped bar.SLDPRT*) downloaded from the publisher's website. This solid model, as shown in Figure 8.3, consists of three solid features (revolve, chamfer, and fillet), a point, and a coordinate system. As soon as you open the model, you may want to check that the IPS unit system is chosen and increase the decimals from the default 2 to 4 digits.

Select NC Machine

Click the CAMWorks feature tree tab 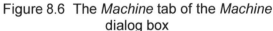. Right click *Machine* and select *Edit Definition*. In the *Machine* dialog box (Figure 8.6), *Mill-in* is selected under *Machine* tab. We choose *Turn Single Turret* from the list of *Applicable machines* box, and click *Select*.

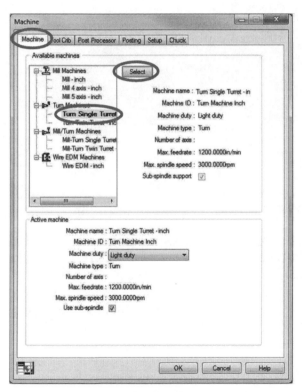

Figure 8.6 The *Machine* tab of the *Machine* dialog box

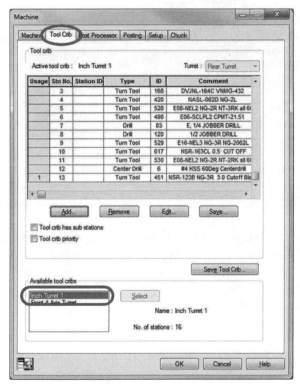

Figure 8.7 The *Tool Crib* tab of the *Machine* dialog box

Choose *Tool Crib* tab and select *Inch Turret 1* under *Available tool cribs* (Figure 8.7), and then click *Select*. We will use the cutters available in *Inch Turret 1* for this lesson. Choose the *Post Processor* tab; a post processor called *T2AXIS-TUTORIAL* is selected by default (Figure 8.8). This is a generic post processor of 2-axis lathe that comes with CAMWorks. There are more post processors that come with CAMWorks; they are located in *CAMWorks2016\Posts* folder in your computer. Note that in practice you will need to identify a suitable post processor that produces G-codes compatible with the CNC lathes at your shop floor. We will not make any change in the post processor selection for this lesson. Then, click the *Setup* tab, and choose *Coordinate System* for *Method* (see Figure 8.9), then click *Coordinate System1* from the graphics area. Click *OK* to accept the selections and close the dialog box.

Create Stock

From CAMWorks feature tree , right click *Stock Manager* and choose *Edit Definition*. The *Stock Manager* dialog box appears (Figure 8.10), in which a default stock size appears under *Bar stock parameters*, including outer diameter ⊟ (4.25in.), inner diameter ⊟ (0in.), overall length ⊟ (6.5in.), and back of stock absolute ⊟ that defines length of the stock outside the part (0in.). The default stock material is *Steel 1005*. We will increase the overall length of the stock to 7.75in. and enter the length of the stock outside the part to be −1.25in., as shown in Figure 8.10. Accept the revised stock by clicking the checkmark ✓ at the top left corner. The bar stock should appear in the graphics area similar to that of Figure 8.4.

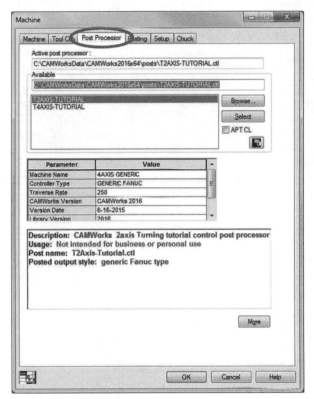

Figure 8.8 The *Post Processor* tab of the *Machine* dialog box

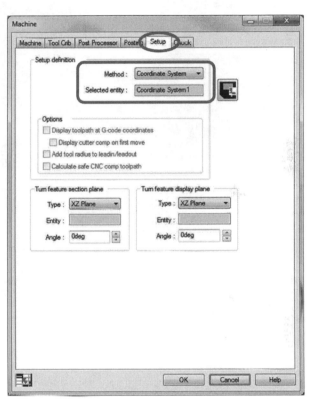

Figure 8.9 The *Setup* tab of the *Machine* dialog box

Turn Setup and Machinable Feature

We create two machinable features manually, OD and cut off. We first create a turn setup and then insert new turn features under the CAMWorks feature tree.

Under the CAMWorks feature tree tab , right click *Stock Manager* and choose *New Turn Setup*. The *Turn Setup* dialog box appears (Figure 8.11). In the graphics area, a coordinate system of the part setup appears at the rear end face of the part (circled in Figure 8.12), coinciding with the fixture coordinate system, *Coordinate System1*. This coordinate system indicates that the Z-axis aligns with the spindle direction, which is adequate. We will relocate the part setup origin to the front end face of the stock later. Click the checkmark ✓ in the *Turn Setup* dialog box to accept the definition. A *Turn Setup1* is now listed in the CAMWorks feature tree .

Now we define machinable features. Under CAMWorks feature tree tab [CW], right click *Turn Setup* and choose *New Turn Feature*.

In the *New Turn Feature* dialog box (Figure 8.13), choose *OD Feature* for *Type*, and pick the five edges that define the boundary profile of the OD feature in the graphics area, as shown in Figure 8.14. Click the checkmark ✓ in the *New Turn Feature* dialog box to accept the definition.

An *OD Feature1* entity is now listed in the CAMWorks feature tree [CW] in magenta color (Figure 8.15).

Figure 8.10 The *Stock Manager* dialog box

Figure 8.11 The *Turn Setup* dialog box

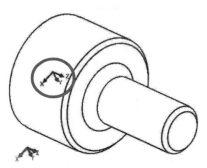

Figure 8.12 The turning coordinate system

Figure 8.13 The *New Turn Feature* dialog box

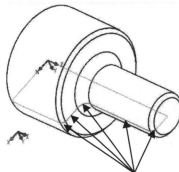

Pick these five edges

Figure 8.14 Picking edges to define an OD feature

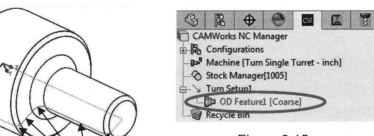

Figure 8.15

We follow the same steps to create a cut off machinable feature; i.e., right click *Stock Manager* and choose *New Turn Feature*. In the *New Turn Feature* dialog box, choose *CutOff Feature* for *Type*, and pick the edge at the rear end face of the part in the graphics area, as shown in Figure 8.16. Click the checkmark ✔ in the *New Turn Feature* dialog box to accept the definition.

A *CutOff Feature1* entity is now listed in the CAMWorks feature tree 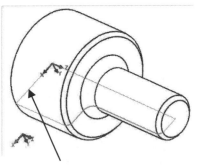 in magenta color.

Pick this edge at the rear end face

Figure 8.16 Picking edge to define a cutoff feature

Generate Operation Plan and Toolpath

Right click *Turn Setup1* and choose *Generate Operation Plan* (or click the *Generate Operation Plan* button above the graphics area).

Three operations: *Turn Rough1*, *Turn Finish1*, and *Cut Off1*, are listed in CAMWorks operation tree in magenta color. Right click *Turn Setup1* and choose *Generate Toolpath* (or click the *Generate Toolpath* button above the graphics area). Turning toolpaths will be generated like those shown in Figure 8.5.

Define Part Setup Origin

A part setup origin has been chosen by CAMWorks to be coincident with the fixture coordinate system, *Coordinate System1* (see Figure 8.16). We will move the part setup origin to *Point1* (located at the center of the front end face of the part), which is more practical since this point is easier to access in a stock clamped into a chuck on a lathe.

Under the CAMWorks operation tree tab , right click *Turn Setup1* and choose *Edit Definition*. In the *Operation Setup Parameters* dialog box, choose *Origin* tab, and pull down the *Defined from* option to choose *Automatic* (Figure 8.17). The part setup origin moved to the center of the front end face (see Figure 8.18) as desired. Click *OK* in the *Operation Setup Parameters* dialog box to accept the change.

Review the Operations

Next we review the tool and key machining parameters of the first turning operation, *Turn Rough1*.

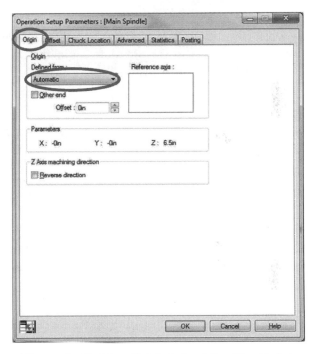

Figure 8.17 The *Origin* tab of the *Tool* tab in the *Operation Setup Parameters* dialog box

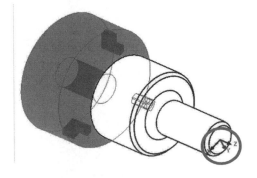

Figure 8.18 The part setup origin moved to the center of front end face

Under the CAMWorks operation tree tab ![icon], right click *Turn Rough1* and choose *Edit Definition*. In the *Operation Parameters* dialog box, choose *Diamond Insert* tab, a diamond insert of 0.0313×80° (radius 0.0313in., angle: 80°, and inscribed circle 0.5in., ID:13, CNMG-432) appears (Figure 8.19). Click the *Holder* tab (see Figure 8.20). A standard holder of shank width 1in. and length 6in. (ID:30, DCLNL-854D-CNMG-432) has been chosen. Also, the *Down left* is chosen for *Orientation*, which defines the orientation of the cutter suitable to turn the OD feature.

Choose the *Turn Rough* tab of the *Operation Parameters* dialog box. In the *Profile parameters* area, the *First cut amt.*, *Max cut amt.* and *Final cut amt.* are set to *0.1in*, as shown in Figure 8.21. These parameters define the distance of the tool movement along the X-direction, similar to the depth of cut in volume milling. Click *Cancel* to close the dialog box.

Simulate Toolpath

Right click *Turn Setup1* and choose *Simulate Toolpath* (or click the *Simulate Toolpath* button ![icon] above the graphics area) to the simulate toolpath. The material removal simulation appears similar to that of Figure 8.1(c).

Step Through Toolpath

Now we take a closer look at the toolpath using the *Step Thru Toolpath* capability to better understand the turning operations generated.

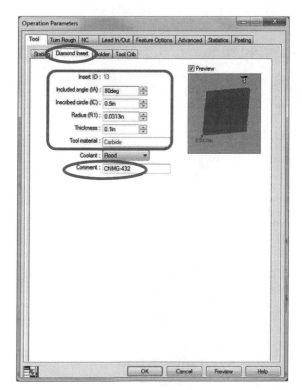

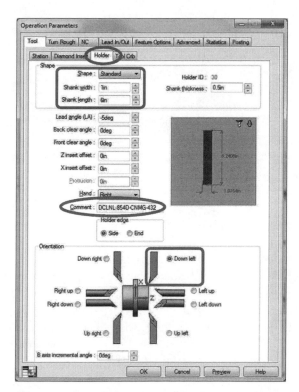

Figure 8.19 The *Diamond Insert* tab of the *Tool* tab in the *Operation Parameters* dialog box

Figure 8.20 The *Holder* tab of the *Tool* tab in the *Operation Parameters* dialog box

Figure 8.21 The *Turn Rough* tab in the *Operation Parameters* dialog box

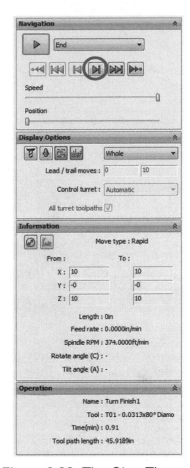

Figure 8.22 The *Step Through Toolpath* dialog box

We pick two operations for a closer look, *Turn Finish* and *Cut Off*, since both are simple and easy to understand. In *Turn Finish* operation, the tool moves along the five edges of the OD machinable feature. In *Cut Off* operation, the tool cuts along the straight edge at the rear end face of the part. Reviewing the toolpath will help us understand the G-codes to be discussed later.

Right click *Turn Finish1* and choose *Step Thru Toolpath*. The *Step Through Toolpath* dialog box appears (Figure 8.22).

Under *Information*, CAMWorks shows the tool movement from the current to the next steps in X, Y, and Z coordinates, the feedrate, spindle speed, among others.

Click the *Step* button 📐 at the center of the *Step Through Toolpath* dialog box (circled in Figure 8.22) to step through the toolpath. You may want to turn on *Show toolpath points* and *Tool Holder Shaded Display* to see the toolpath display similar to that of Figure 8.23.

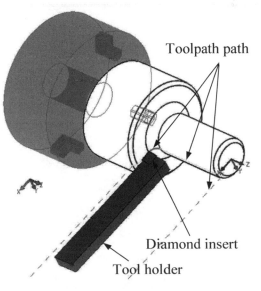

Figure 8.23 Stepping through the toolpath

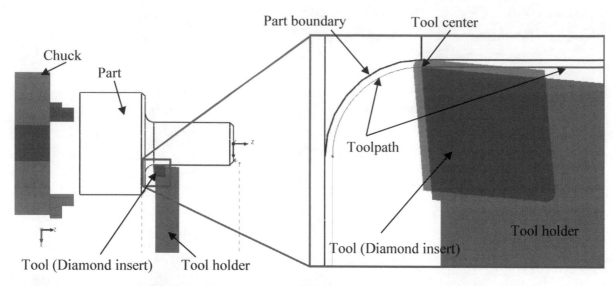

Figure 8.24 Toolpath of *Turn Finish*, a closer look

A closer look at the toolpath of *Turn Finish* (see Figure 8.24) indicates that the toolpath follows the trace of the tool center (i.e., the center of the corner radius of the cutter). As a result, in general, the toolpath offsets an amount of tool radius from the part boundary. The XZ coordinates of the six characteristic points of the part boundary, A to F shown in Figure 8.25, referring to the part setup origin are listed in Table 8.1. The toolpath offset by an amount of the tool radius, 0.0313in., is especially clear at points B, C, and D.

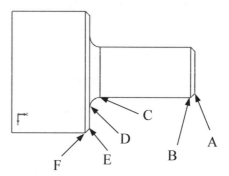

Figure 8.25 The six characteristic points (A to F) of the part boundary

Table 8.1 The XZ Coordinates of the Six Characteristic Points A to F

Point	X and Z Coordinates of Part Boundary	X and Z Coordinates of Toolpath	Offsets in X and Z Coordinates
A	X = 0.725 Z = 0	X = 0.718 Z = 0.0575	$\Delta X = 0.007$ $\Delta Z = 0.0575$
B	X = 0.875 Z = −0.15	X = 0.9063 Z = −0.137	$\Delta X = 0.0313$ $\Delta Z = 0.0313$
C	X = 0.875 Z = −3.375	X = 0.9063 Z = −3.375	$\Delta X = 0.0313$ $\Delta Z = 0$
D	X = 1.25 Z = −3.75	X = 1.25 Z = −3.7187	$\Delta X = 0$ $\Delta Z = 0.0313$
E	X = 1.975 Z = −3.75	X = 1.988 Z = −3.7187	$\Delta X = 0.013$ $\Delta Z = 0.0313$
F	X = 2.125 Z = −3.9	X = 2.1471 Z = −3.8779	$\Delta X = 0.0221$ $\Delta Z = 0.0221$

Now we follow the same steps to review the toolpath of the *Cut Off* operation.

Right click *Cut Off1* and choose *Step Thru Toolpath*. Note that the Z coordinates shown in the *Step Through Toolpath* dialog box (Figure 8.26) are −6.5625, while the rear end face of the part is 6.5in. to the left of the part setup origin (again, located at the center of the front end face of the part). This is because a groove cutter of 0.125in. in width was selected by CAMWorks for the operation. The right edge of the tool is in contact with the part boundary, as seen Figure 8.27. Therefore, the toolpath is offset an amount of half cutter width to the left of the part boundary. That is, the Z-coordinates of the toolpath is −6.5−0.125/2 = −6.5625, as expected.

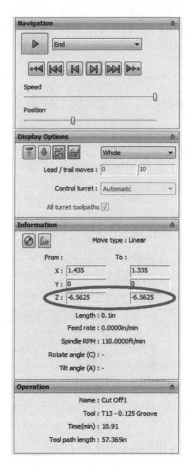

Figure 8.26 The *Step Through Toolpath* dialog box

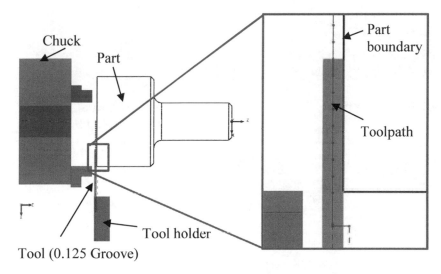

Figure 8.27 Toolpath of *Cut Off*, a closer look

8.4 Reviewing the Turn G-codes

We are now ready to generate and review G-codes for *Turn Finish* and *Cut Off* operations.

Right click *Turn Finish1* and choose *Post Process*. In the *Post Output File* dialog box (Figure 8.28), choose a proper file folder, enter a file name (for example, *Turn finish.txt*). The *Post Process Output* dialog box appears (Figure 8.29).

In the *Post Process Output* dialog box, click the *Play* button ▶ (circled in Figure 8.29) to create G-code (.txt file).

Open the *Turn finish.txt* file from the folder using *Word* or *Word Pad*; see the file contents shown in Figure 8.30(a). It is shown that the NC blocks N9 and N10 moves the cutter to Point A, N11 to Point B, N12 to Point C, N13 to Point D, N14 to Point E, and N15 to Point F.

Repeat the same for the cut off operation. The NC block N9 shows the Z-coordinate of the toolpath; see Figure 8.30(b). Note that the Z-coordinate shown in the G-code is −6.625, which should have been

−6.5625 as indicated in Figure 8.26. This is apparently an error in the post processor of CAMWorks and must be corrected before implementing the G-code on a lathe.

We have now completed the lesson. You may save your model for future reference.

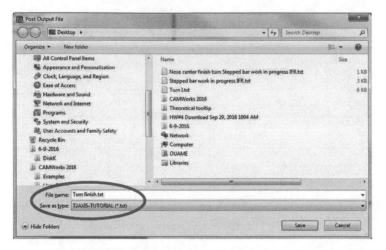

Figure 8.28 The *Post Output File* dialog box

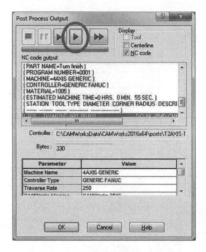

Figure 8.29 The *Post Process Output* dialog box

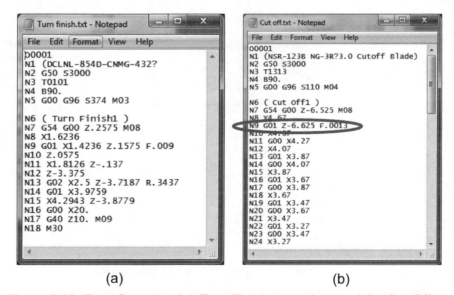

(a)   (b)

Figure 8.30 Turn G-codes, (a) *Turn Finish* operation, and (b) *Cut Off* operation

8.5 Exercises

Problem 8.1. Repeat the same steps in this lesson to generate turning operations, except that we use the automatic feature recognition capability by clicking the *Extract Machinable Features* button [Extract Machinable Features] above the graphics area. Is there any redundant machinable feature extracted? Are there any noticeable differences in the toolpath of the three turning operations, turn rough, turn finish, and cut off, compared with those discussed in this lesson?

Lesson 9:   Turning a Stub Shaft

9.1   Overview of the Lesson

In Lesson 9, we machine a stub shaft, which is similar to that of Lesson 8 but with more turning features, including face, groove, thread, and holes at both ends. Since the two holes are located at the front and rear ends, respectively, there will be two turn setups that support turning operations to machine the features from both ends of the bar stock.

Lesson 9 offers a more in-depth discussion in creating turning operations using CAMWorks. You will learn to use automatic feature recognition (AFR) to extract machinable features and make necessary adjustments for a valid turning simulation that can be implemented physically. We will use the same machine as Lesson 8, that is, a lathe of single turret, to machine the stub shaft example shown in Figure 9.1(a) from a bar stock clamped into a three-jar chuck from its rear end, then to the front end, to turn machinable features from both ends—only the chuck at the rear end is shown in Figure 9.1(b). Therefore, there are at least two turn setups required for this example. The machined part at the end of the material removal simulation of the first turn setup (*Turn Setup1*) is shown in Figure 9.1(c).

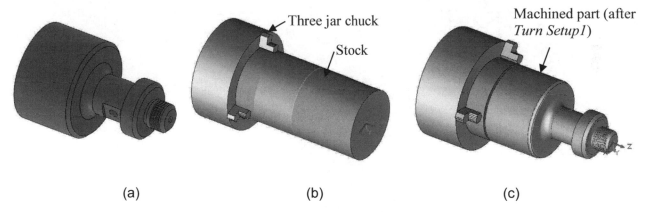

(a)   (b)   (c)

Figure 9.1 The stub shaft example, (a) the design model, (b) bar stock clamped to a chuck, and (c)
the material removal simulation (after *Turn Setup1*)

Most of the turning operations generated by CAMWorks are adequate for this example, except for the threading operation. Although we stay with default options and parameters for most of the lesson, we point out a few deficiencies that need to be corrected, including the threading operation. We will take a closer look at the threading operation by stepping through the toolpath and reviewing the G-code to gain a better understanding. At the end we briefly introduce the procedure of using a turn-mill to machine the two side cuts and a cross hole near the mid portion of the shaft. These features will have to be taken care of using a mill instead of a lathe. A turn-mill is suitable to machine such a part with both turn and mill machinable features.

After completing this lesson, you should be able to carry out a turning simulation for applications of most turning features following the same procedures.

9.2   The Stub shaft Example

Similar to the stepped bar example, the stub shaft has a bounding cylinder of size ϕ4.25in.×6.5in. In addition to the revolve, chamfer and fillet features, there are side cuts, a cross hole, and holes at the two ends; see Figure 9.2(a). Note that the sketch of the revolve feature shown in Figure 9.2(b) is a bit more complex than that of Lesson 8, which leads to additional machinable features, such as a groove feature. The sketches of the end holes and thread are shown in Figure 9.2 (c). The thread pitch is 0.55in. Note that the reference features, including a point and a coordinate system, are identical to those of Lesson 8. These features are listed in the feature tree shown in Figure 9.3. Similar to Lesson 8, the coordinate system (*Coordinate System1*) is chosen as the fixture coordinate system. Both the origin of *Coordinate System1* and *Point1* are defined as the part setup origins, respectively, for the two turn setups. When you open the solid model *Stub shaft.SLDPRT*, you should see the solid features and the reference features listed in the feature tree (see Figure 9.3).

The unit system chosen is IPS.

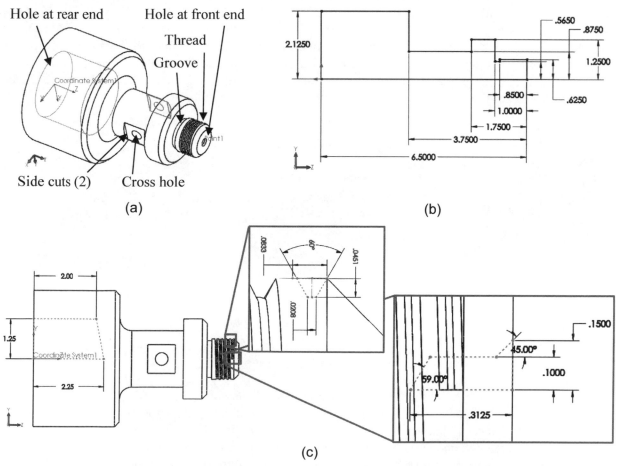

Figure 9.2 The stub shaft solid model, (a) major solid features, (b) sketch and dimensions of the revolve solid feature, and (c) sketches and dimensions of the two end holes and thread

A bar stock of size φ4.25in.×8in., made of low carbon alloy steel (1005), as shown in Figure 9.4(a), is chosen for the machining operations. The front end of the stock is extended by 0.25in. from the front face of the part. The rear end of the stock is fixed in the three-jar chuck with an extra length of 1.25in. to be cut off at the end of the turning operations (similar to that of Lesson 8) of the first turn setup. Due to the extra material at the front end face, a face turning operation is required.

Note that the part setup origins are defined at the center of the front end face of the part (coincides with *Point1* defined in part) and the origin of the coordinate system, *Coordinate System1*, for the two turn setups, respectively. The first setup cuts all features except for the hole at the rear end, which is machined by the operations of the second setup.

We first extract machinable features using the automatic feature recognition (AFR) capability. Most machinable features are extracted, including face, OD, grooves, ID (the two holes). The only feature that is not extracted is the thread. We manually create a thread machinable feature. Overall, there are eight machinable features included, as shown in Figure 9.5. In this lesson, we follow the recommendations of the technology database (TechDB™) for determining machining options, tools, and cutting parameters.

The side cuts and the cross hole are to be machined by using a turn-mill. This topic will be discussed at the end of the lesson.

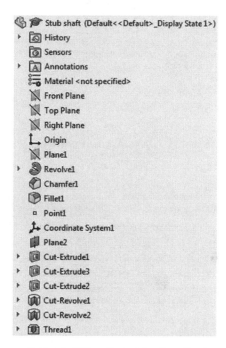

Figure 9.3 Entities listed in the feature tree

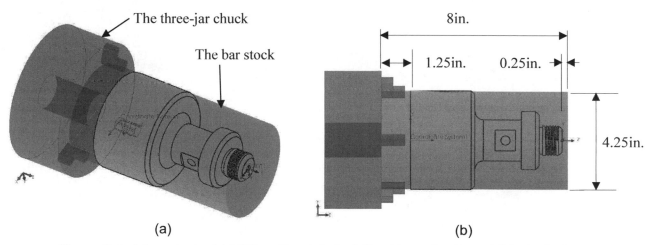

(a)   (b)

Figure 9.4 A bar stock of φ4.25in.×8in., (a) stock fixed in a chuck, and (b) stock size

The toolpath of the turn operations for machining the face, OD, and groove features usually consists of rough and finish operations. Hole drilling consists of center drill and drill operations. The rear end hole requires additional operations, such as bore, mainly due to its size. There are a total of seventeen operations created to machine this stub shaft. Toolpaths of the seventeen operations are shown in Figure 9.6. You may open the example file, *Stub shaft with toolpath.SLDPRT*, to preview the toolpath and turn operations created for this example.

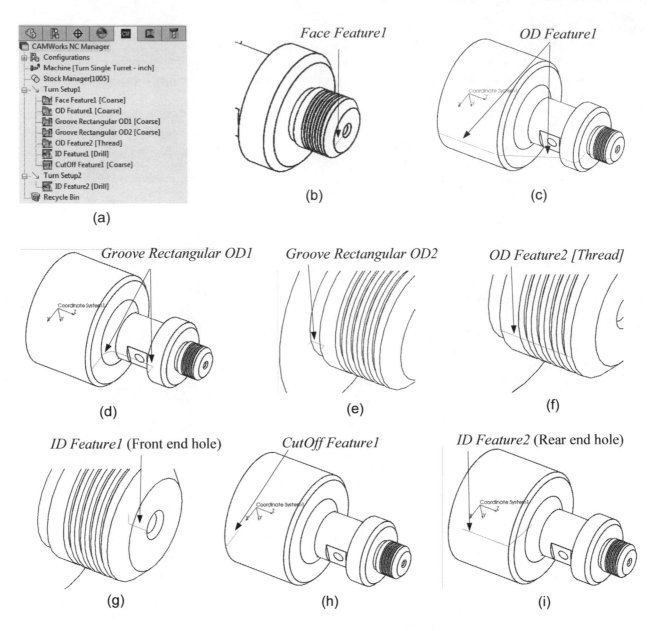

Figure 9.5 The machinable features of the turning operations: (a) CAMWorks feature tree, (b) *Face Feature1*, (c) *OD Feature1*, (d) *Groove Rectangular OD1*, (e) *Groove Rectangular OD2*, (f) *OD Feature2* (thread), (g) *ID Feature1* (front end hole), (h) *CutOff Feature1*, and (i) *ID Feature2* (rear end hole) of *Setup2*

9.3   Using CAMWorks

Open SOLIDWORKS Part

Open the stub shaft part (filename: *Stub shaft.SLDPRT*) downloaded from the publisher's website. This solid model shown in Figure 9.2 consists of nine solid features (revolve, chamfer, fillet, 5 cut extrudes and a thread), a point, two planes, and a coordinate system (see the solid feature tree in Figure 9.3). As soon as you open the model, you may want to check the unit system and increase the decimals from the default 2 to 4 digits.

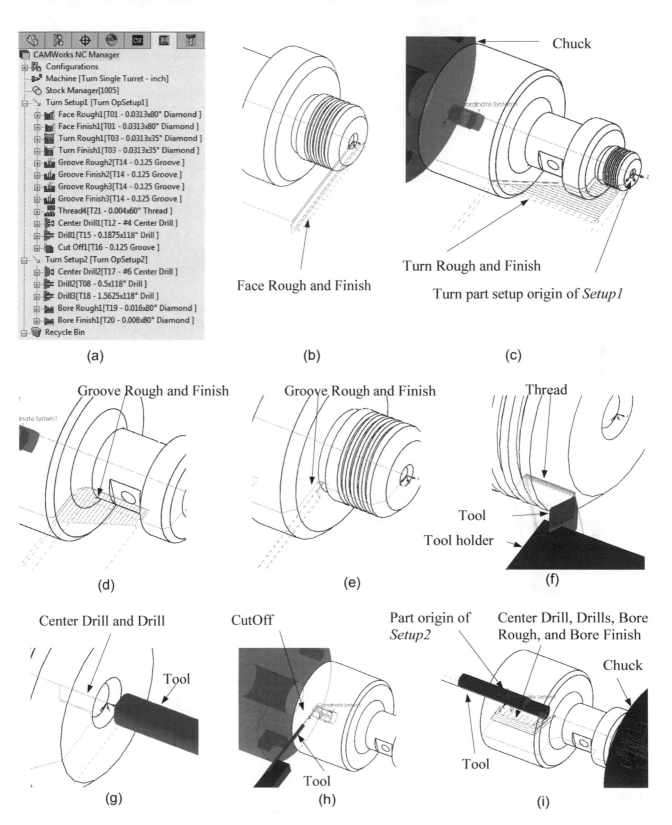

Figure 9.6 The toolpaths of the turning operations, (a) CAMWorks operation tree, (b) *Face Rough1* and *Face Finish1*, (c) *Turn Rough1* and *Turn Finish1*, (d) *Groove Rough2* and *Groove Finish2*, (e) *Groove Rough3* and *Groove Finish3*, (f) *Thread4*, (g) *Center Drill1* and *Drill1*, (h) *Cut Off1*, and (i) *Center Drill2*, *Drill2*, *Drill3*, *Bore Rough1*, and *Bore Finish1* of *Setup2*

Select NC Machine

Click the CAMWorks feature tree tab [CW]. Right click *Machine* and select *Edit Definition*. Similar to those of Lesson 8, under *Machine* tab of the *Machine* dialog box, we select *Turn Single Turret* from the list of *Applicable machines* box, and click *Select*. We choose *Crib1* under *Available tool cribs* of the *Tool Crib* tab, select *T2AXIS-TUTORIAL* under the *Post Processor* tab, and select *Coordinate System1* under the *Setup* tab.

Create Stock

From CAMWorks feature tree [CW], right click *Stock Manager* and choose *Edit Definition*. In the *Stock Manager* dialog box (Figure 9.7), we enter 8in. for the stock length, increasing by 0.25in. to the right of the part from that of Lesson 8. The default stock material is *Steel 1005*. We enter the length of the stock outside the part to be −1.25in., as shown in Figure 9.7. As a result, a 0.25in. of extra material appears to the right end face of the part. Accept the stock by clicking the checkmark ✔ at the top left corner. The bar stock should appear in the graphics area similar to that of Figure 9.4.

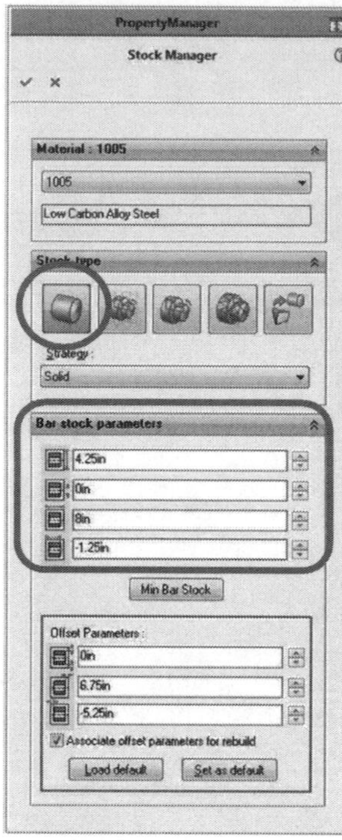

Figure 9.7 The *Stock Manager* dialog box

Turn Setup and Machinable Feature

Click the *Extract Machinable Features* button above the graphics area. Two setups, *Turn Setup1* and *Turn Setup2*, are created with multiple machinable features extracted. Under *Setup1*, there are six features extracted: *Face Feature1*, *OD Feature1*, *Groove Rectangular OD1*, *Groove Rectangular OD2*, *ID Feature1* and *CutOff Feature1*. Under *Setup2*, there is one feature extracted, *ID Feature2*. All machinable features are in magenta color, as shown in Figure 9.8. If you click any of these features, you should see them in the graphics area (as lines or curves) like those shown in Figure 9.5.

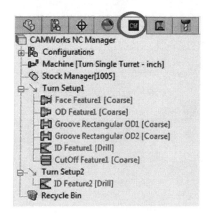

Figure 9.8 The machinable features extracted

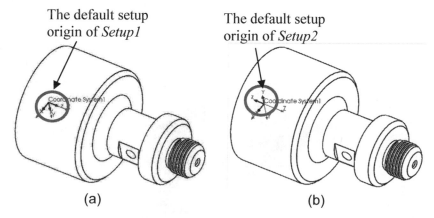

Figure 9.9 The default part setup origins of *Setup1* and *Setup2* at the origin of *Coordinate System1*

Click *Turn Setup1* and *Turn Setup2* to locate the respective setup origins in the graphics area. As shown in Figure 9.9, both origins are located at the origin of *Coordinate System1* with Z-axis points in opposite directions. The directions are all good. We will move the origin of *Setup1* to the front end face of the part (coinciding with *Point1*), and keep the origin of *Setup2* as it is. Before we make this change, we will generate operation plans first.

Generate Operation Plan and Toolpath

Right click *Turn Setup1* and choose *Generate Operation Plan* (or click the *Generate Operation Plan* button ⬛ above the graphics area).

Thirteen and five operations are generated for *Turn Setup1* and *Turn Setup2*, respectively (Figure 9.10). All are in magenta color.

Click the *Generate Toolpath* button 🔧 above the graphics area to generate toolpath. Turning toolpaths are generated for all operations, except for *Groove Rough1* and *Groove Finish1*, which remain in magenta color, as shown in Figure 9.11. Note that *Groove Rough1* and *Groove Finish1* operations are generated for *OD Feature1* feature. In fact, *Turn Rough1* and *Turn Finish1* operations are turning *OD Feature1* already. Why are there operations to machine the same feature and yet the toolpaths are not generated?

This is because the thread solid feature at the front end of the shaft was not correctly extracted. The *Groove Rough1* and *Groove Finish1* operations are generated to machine the thread without accurate geometric information of the machinable feature. You may simply delete these operations by right clicking them and choose *Delete*. The deleted entities are moved under *Recycle Bin* in the operation tree.

Figure 9.10 The operations generated

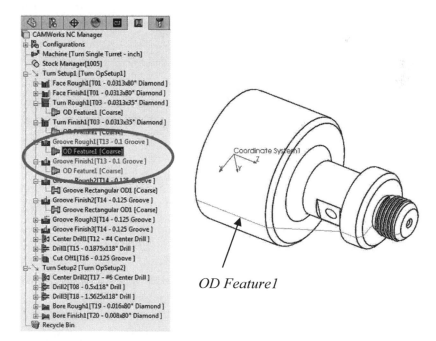

OD Feature1

Figure 9.11 Toolpath not generated for *Groove Rough1* and *Groove Finish1* operations

Relocate Part Setup Origin of *Setup1*

As mentioned earlier, we will move the origin of *Setup1* to the front end face of the part (coinciding with *Point1*, located at the center of the front end face of the part), which is more practical since this point is easier to access in setting up the stock on a lathe.

Under the CAMWorks operation tree tab ▣ , right click *Turn Setup1* and choose *Edit Definition*. In the *Operation Setup Parameters* dialog box, choose *Origin* tab, and pull down the *Defined from* option to choose *Automatic* (Figure 9.12). Click *Yes* to the question in the *CAMWorks Warning* box: *The origin has changed, toolpaths need to be recalculated. Genenerate toolpath now?*

The part setup origin is now moved to the center of front end face (see Figure 9.13) as desired. Click *OK* in the *Operation Setup Parameters* dialog box to accept the change.

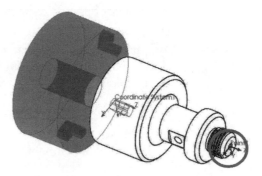

Figure 9.13 The part setup origin moved
to the center of front end face

Figure 9.12 The *Origin* tab of the *Operation
Setup Parameters* dialog box

Review the Operations

Next we review the tool and key machining parameters of a selected operation, *Groove Rough1*.

Under the CAMWorks operation tree tab ▣ , right click *Groove Rough1* and choose *Edit Definition*. In the *Operation Parameters* dialog box, choose *Tool* tab and then *Groove Insert* tab, a groove insert of 0.125in. in width (Insert ID: 80, NG-3125L) appears (Figure 9.14). Tool holder with groove insert appears in the graphics area (Figure 9.15). Click the *Holder* tab (see Figure 9.16); a standard holder of shank width 0.375in. and length 4.5in. has been chosen. Also, the *Down left* is chosen for *Orientation*, which defines the orientation of the cutter suitable to turn the groove feature.

Choose the *Groove Rough* tab of the *Operation Parameters* dialog box. In the *Parameters* area, the *Stepover* is set to *0.1in*, and *Allowance* is set to *0.01in* for both X and Z directions, as shown in Figure 9.17.

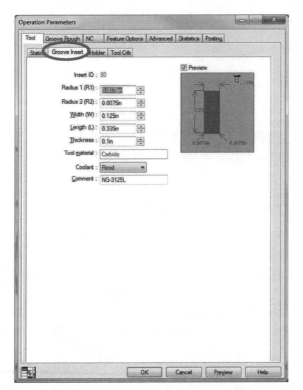

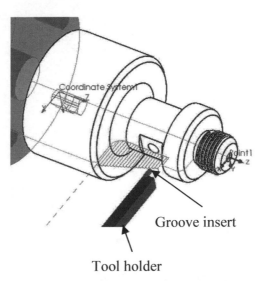

Figure 9.15 Tool holder with groove insert appears in the graphics area

Figure 9.14 The *Groove Insert* tab of the *Tool* tab of the *Operation Parameters* dialog box

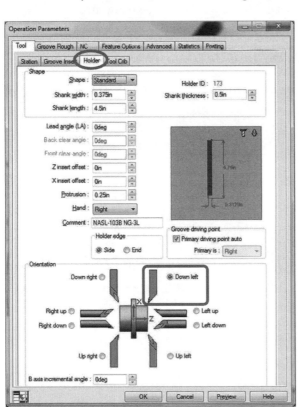

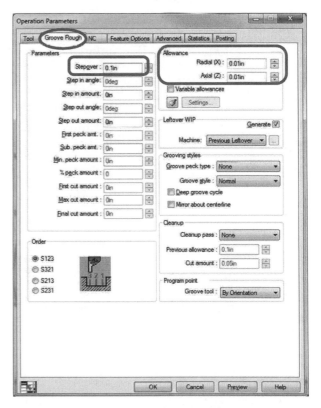

Figure 9.16 The *Holder* tab of the *Tool* tab of the *Operation Parameters* dialog box

Figure 9.17 The *Groove Rough* tab of the *Operation Parameters* dialog box

The *Stepover* defines the distance of the tool movement along the X-direction, similar to the depth of cut in milling operation. Allowances leave a small amount of material uncut, similar to those of rough mill operations. Click *Cancel* to close the dialog box since we are not making any changes to the *Groove Rough1* operation.

Simulate Toolpath

Click the *Simulate Toolpath* button above the graphics area) to simulate toolpath. The material removal simulation appears similar to those of Figure 9.18(a) and Figure 9.18(b) for *Turn Setup1* and *Turn Setup2*, respectively.

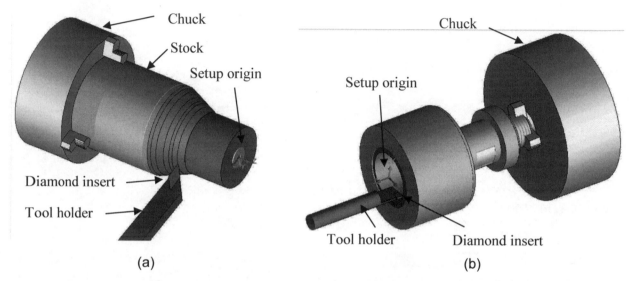

(a)   (b)

Figure 9.18 The material removal simulation, (a) *Turn Setup1*, and (2) *Turn Setup2*

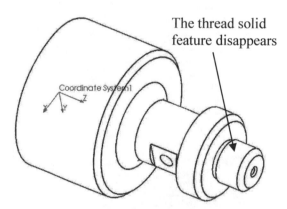

Figure 9.19 The thread solid feature suppressed

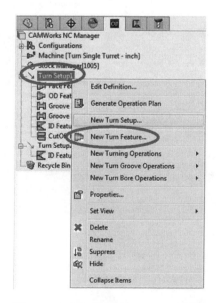

Figure 9.20 Creating a threading machinable feature

9.4   Cutting the Thread

We now take a look at the thread. Since the thread solid feature was not extracted as a machinable feature, we will have to manually create one. First, we suppress the thread solid feature by clicking the *FeatureManager* design tree tab 🔑, right clicking *Thread1* and choosing *Suppress*. The thread solid feature disappears in the part in the graphics area (see Figure 9.19).

From CAMWorks feature tree 🅲🆆, right click *Turn Setup1* and choose *New Turn Feature* (Figure 9.20).

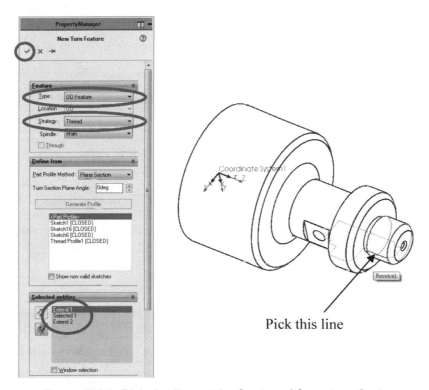

Pick this line

Figure 9.21 Pick the line at the front end for a turn feature

In the *New Turn Feature* dialog box (Figure 9.21), choose *OD Feature* for *Type* (should have been chosen by default), select *Thread* for *Strategy*, and then pick the line segment near the front end of the part in the graphics area (Figure 9.21). The line picked is now listed under *Selected Entities*. Click the checkmark to accept the turn feature.

In the CAMWorks feature tree 🅲🆆, an *OD Feature2 [Thread]* is added and is in magenta color under *Turn Setup1*, as shown in Figure 9.22.

Right click *OD Feature2 [Thread]* and choose *Generate Operation Plan*. A *Thread1* operation is now added to the operation tree under *Turn Setup1*, again in magenta color.

Right click *Thread1* and choose *Generate Toolpath*. Toolpath will be generated similar to that of Figure 9.6(f).

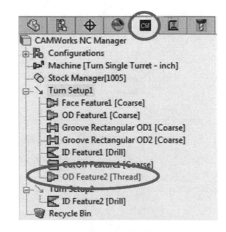

Figure 9.22

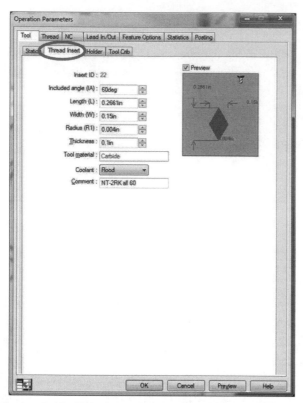

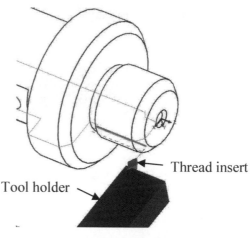

Figure 9.24 Tool holder with thread insert appears in the graphics area

Figure 9.23 The *Thread Insert* tab of the *Tool* tab of the *Operation Parameters* dialog box

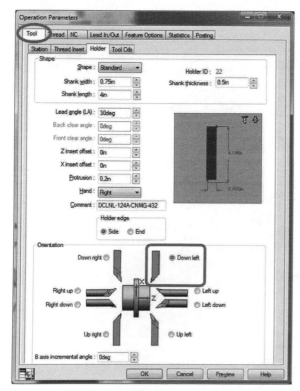

Figure 9.25 The *Holder* tab of the *Tool* tab of the *Operation Parameters* dialog box

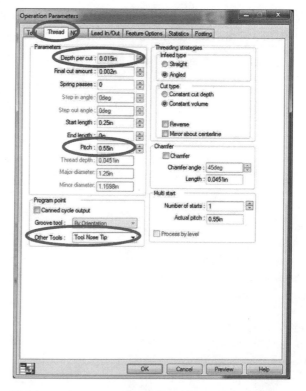

Figure 9.26 The *Thread* tab of the *Operation Parameters* dialog box

Next we take a closer look at the thread operation and thereafter, the toolpath.

Under the CAMWorks operation tree tab , right click *Thread1* and choose *Edit Definition*. In the *Operation Parameters* dialog box, choose *Tool* tab and then *Thread Insert* tab, a thread insert of 0.004×60° (radius 0.004in. included angle 60 degrees, thickness 0.1in.) appears (Figure 9.23). Tool holder with thread insert appears in the graphics area (Figure 9.24). Click the *Holder* tab (see Figure 9.25); a standard holder of shank width 0.75in. and length 4in. has been chosen. Also, the *Down left* is chosen for *Orientation*, which defines the orientation of the cutter suitable to turn the thread feature.

Click the *Thread* tab, and note that the *Depth per cut* is *0.015in.*, *Pitch* is *0.0625in.* (which is inconsistent with that of the solid feature and will be corrected next), and *Tool Nose Tip* is chosen for *Program point*. The other option for *Program point* is *Tool Nose Center*. Different options affect the toolpath and G-codes, which will be discussed later in this lesson.

Click the *Feature Options* tab, and click *Parameters* (see Figure 9.27). In the *OD Profile Parameters* dialog box (Figure 9.28), enter *1.25in.* for *Major dia.*, *0.0451in.* for *Thread depth*, and *0.55in.* for *Pitch* so as to make the thread size consistent with that of the solid feature (see dimensions of the thread in Figure 9.2(c)). Click *OK* in the *OD Profile Parameters* dialog box. The *Pitch* is now *0.55in.* under the *Thread* tab of the *Operation Parameters* dialog box. Click *OK* in the *Operation Parameters* dialog box to accept the changes.

Figure 9.27 The *Feature Options* of the *Operation Parameters* dialog box

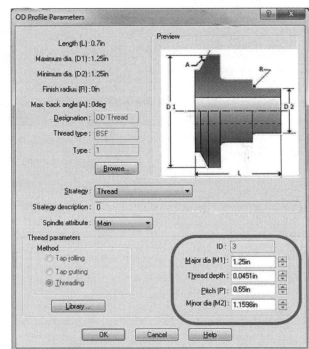

Figure 9.28 The *OD Profile Parameters* dialog box

Step Through Toolpath

Now we take a closer look at the toolpath to better understand the toolpath (and later the G-code) of the threading operation.

Right click *Thread1* and choose *Step Thru Toolpath*. The *Step Through Toolpath* dialog box appears (Figure 9.29).

Under *Information*, CAMWorks shows the tool movement from the current to the next steps, the feedrate, and spindle speed, among others.

Click the *Step* button [▶] at the center (circled in Figure 9.29) to step through the toolpath. You may want to turn on *Show toolpath points* and *Tool Holder Shaded Display* to see the toolpath display similar to that of Figure 9.30.

A zoom in view of the toolpath (see Figure 9.30) indicates that the toolpath follows the trace of the tool center (center of the corner radius of the insert). As a result, in general, the toolpath offsets an amount of tool radius from the part boundary. For example, the XZ coordinates of the last cutting pass, as shown in Figure 9.31(a), are from (0.5799, 0.1) to (0.5799, −0.85).

The X coordinates indicate that the last pass is 0.5799in. away from the part set up origin, as shown in Figure 9.31(a), which is *Point1* (center of the front end face of the part). The radius of the revolve feature where thread is created is 0.625in.; see Figure 9.2(b). The thread depth is 0.0451in.; see Figure 9.2(c). Therefore, the X coordinate of the last cutting pass is 0.625−0.0451 = 0.5799 as it should be.

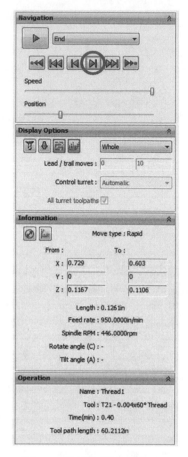

Figure 9.29 The *Step Through Toolpath* dialog box

Recall that we chose *Tool Nose Tip* for *Program point*; hence, XYZ coordinates of the toolpath are generated at the tool tip. If you choose *Tool Nose Center* for *Program point*, the X coordinate of the last cutting pass will be 0.625−0.0451+0.004 = 0.5839, in which 0.004in. of the tool radius is added to correctly offset the toolpath.

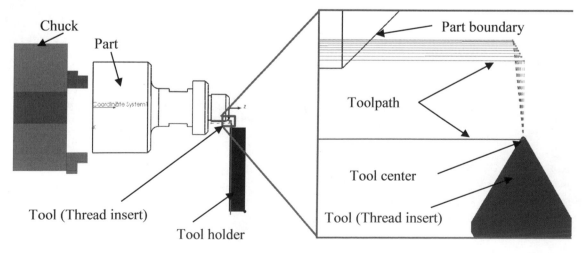

Figure 9.30 Toolpath of *Thread1*, a closer look

G-Code of the Threading Operation

Right click *Thread1* and choose *Post Process* to generate the G-code.

Open the *Stub shaft.txt* file (see the file contents shown in Figure 9.32). It is shown that the NC blocks N45 and N45 move the cutter along the last cutting pass, in which the X-coordinate is 1.1598 (= 2×0.5799), which is correctly converted.

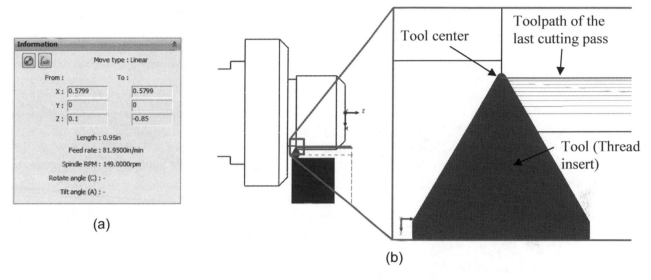

(a)

(b)

Figure 9.31 The last cutting pass of the toolpath of *Thread1*, (a) the XZ coordinates of the toolpath, and (b) a zoom in view of the cutter location

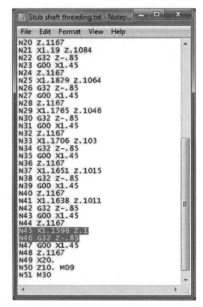

Figure 9.32 The turn G-code of *Thread1* operation

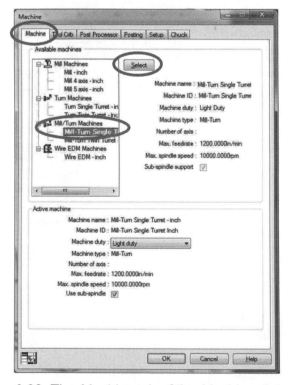

Figure 9.33 The *Machine* tab of the *Machine* dialog box

We have completed the first part of the exercise. You may want to save your model. Next we will start over using a turn-mill machine that is able to cut both turn and mill machinable features.

9.5 Using a Turn-Mill

We first suppress the thread solid feature to avoid the problem encountered earlier; that is, the thread solid feature was not recognized as a machinable feature (right click *Thread1* and choose *Suppress*).

Select NC Machine

Click the CAMWorks feature tree tab ▣. Right click *Machine* and select *Edit Definition*. In the *Machine* dialog box (Figure 9.33), choose *Mill-Turn Single Turret* from the list of *Applicable machines* box, and click *Select*.

Choose *Tool Crib* tab, select *MT Crib 1*, and then click *Select*. Choose the *Post Processor* tab; a post processor called *MT2AXIS-TUTORIAL* is selected. This is a generic post processor of 2-axis mill-turn machine that comes with CAMWorks. Then, click the *Setup* tab, and choose *Coordinate System* for *Method*, then click *Coordinate System1* from the graphics area. Click *OK* to accept the selections and close the dialog box.

Create an identical bar stock; i.e., with outer diameter (4.25in.), inner diameter (0in.), overall length (8in.), length of the stock outside the part (−1.25in.), and default stock material *Steel 1005*.

Extract Machinable Features

Click the *Extract Machinable Features* button ▣ above the graphics area. In addition to *Turn Setup1* and *Turn Setup2*, two more setups are generated, *Mill Setup1* and *Mill Setup3*.

Under *Mill Setup1*, there is one feature extracted, *Hole1*; and Under *Mill Setup3*, there are two features, *Irregular Slot1* and *Irregular Slot2*, as shown in Figure 9.34. Machinable features extracted are also listed under *Turn Setup1* and *Turn Setup2*. We will focus on the mill operations and leave turn operations as they are since turn operations have been discussed in detailed.

Figure 9.34 The machinable
features extracted

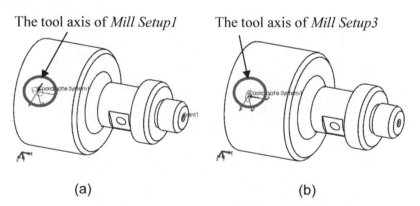

(a)   (b)

Figure 9.35 The tool axes of (a) *Mill Setup1* and (b) *Mill Setup3*

We first check the tool axes of the respective two mill setups. Click *Mill Setup1* under the feature tree. A tool axis symbol ⚒ appears at the origin of the fixture coordinate system, *Coordinate System1*, with the arrow pointing in the X-direction of the fixture coordinate system, as shown in Figure 9.35(a), indicating the tool axis of the mill setup. Since the cross hole is the only machinable feature extracted under *Mill Setup1*, the tool axis of the setup is adequate. No change is necessary.

Next, we click *Mill Setup3*. A tool axis symbol ⚒ appears again at the origin of the fixture coordinate system, but with the arrow pointing in the negative Z-direction of the fixture coordinate system; see Figure 9.35(b). The two machinable features under this setup are the two side cuts that must be machined by tools with the axis pointing in the X-direction. It is apparent that the tool axis of *Mill Setup3* is incorrect and must be changed.

Also, the two side cuts are located on the opposite sides of the shaft. They will have to be machined by operations under the two mill setups, respectively.

We first give it a try to modify the tool axis of *Mill Setup3* by right clicking it and choosing *Edit Definition*. In the *Mill Setup* dialog box (Figure 9.36), a face is selected, and in the graphics area, a tool axis symbol appears like that of Figure 9.35(b).

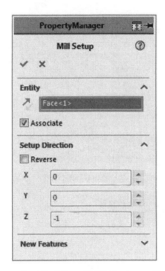

Figure 9.36 The *Mill Setup* dialog box

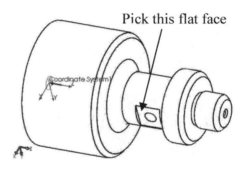

Figure 9.37 Picking the face of the side cut
at the front side (*Irregular Slot1*)

Pick the flat face of the side cut at the front side (*Irregular Slot1*), as shown in Figure 9.37. A CAMWorks message appears: *direction does not match selected Part Setup direction*. It indicates that changing the tool axis for *Mill Setup3* is not possible in CAMWorks. Click *OK* to close the message box.

We will create a mill setup manually by right clicking *Stock Manager* and choosing *New Mill Part Setup* (under CAMWorks feature tree 🔲). The same *Mill Setup* dialog box appears. We now pick the flat face of the side cut at the front side, as shown in Figure 9.37. A tool axis symbol appears ⚒ at the fixture coordinate system, pointing in the negative X-direction (Figure 9.38), which is adequate. Click the checkmark to accept the setup. A *Mill Setup4* appears in the feature tree.

You may extract machinable features for *Mill Setup4* by right clicking it and choosing *Recognize Features*. A machinable feature *Rectangular Slot1* is extracted and the side cut at the front is highlighted in the part (like that of Figure 9.39), indicating that the desired feature is extracted.

Right click *Rectangular Slot1* and choose *Generate Operation Plan* and then *Generate Toolpath*. Two operations, *Rough Mill1* and *Contour Mill1* are generated, as shown in Figure 9.40(a) and Figure 9.40(b), respectively.

You may show material removal simulation by first suppressing *Mill Setup3* and click the *Simulate Toolpath* [Simulate Toolpath] button above the graphics area. The simulation should appear like that of Figure 9.40(c).

Now we do the same to extract the side cut feature at the rear side by right clicking *Mill Setup1* and choosing *Extract Features*. The cut at the rear side is extracted as *Rectangular Slot2*. As a result, there are two machinable features listed under *Mill Setup1*, including *Hole1* extracted earlier.

Generate operation plans and toolpaths for both machinable features. Four operations are created: *Center Drill2* and *Drill2* for *Hole1*, and *Rough Mill2* and *Contour Mill2* for *Rectangular Slot2*.

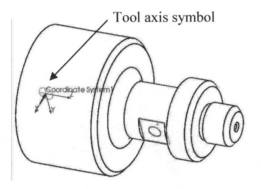

Figure 9.38 Tool axis symbol appears
at the fixture coordinate system,
pointing in the negative X-direction

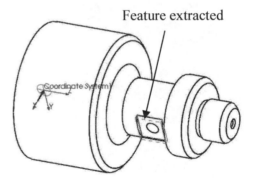

Figure 9.39 The machinable feature
extracted

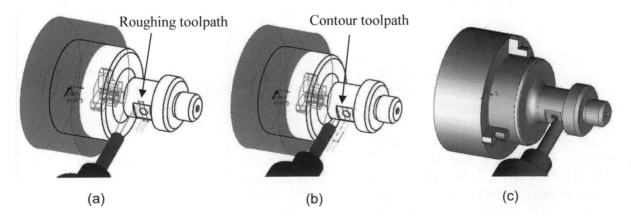

(a)   (b)   (c)

Figure 9.40 The toolpath of the operations under *Mill Setup4*, (a) *Rough Mill1*, (b) *Contour Mill1*, and
(c) material removal simulation

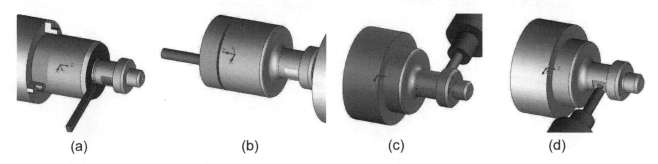

| (a) | (b) | (c) | (d) |

Figure 9.41 The material removal simulation, (a) *Turn Setup1*, (b) *Turn Setup2*, (c) *Mill Setup1*, and (d) *Mill Setup4*

You may create a material removal simulation to simulate the entire machining operation, including turning and milling, as shown in Figure 9.41. Note that the chuck is slightly misplaced in the material removal simulation of *Mill Setup1* and *Mill Setup4*. You may delete *Mill Setup3*.

We have now completed the exercise. You may save your model for future reference.

9.6   Exercises

Problem 9.1. Generate NC operations to machine the part shown in Figure 9.42 from a stock of $\phi2.75$in.$\times$4in. Pick adequate tools (with justifications). Please submit the following for grading:

(a)   A summary of the turn operations, including cutting parameters and tools selected.
(b)   A summary of the mill operations, including cutting parameters and tools selected.
(c)   Screen shots of combined NC toolpaths and material removal simulations.
(d)   Is there any material remaining uncut? If so, was it part design issue or machining issue? What can be done to improve either the part design or machining operations?

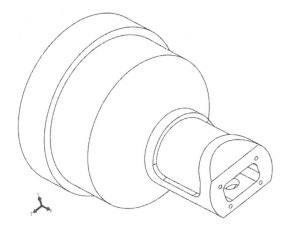

Figure 9.42 The design model of Problem 9.1

Lesson 10: Die Machining Application

10.1   Overview of the Lesson

In this lesson, we present an industrial application that involves die machining of tooling manufacturing for sheet metal forming. In this application, manufacturing of tooling for sheet metal forming, including punch and die, was carried out by using a HAAS mill. CAMWorks was employed to conduct virtual machining and toolpath generation for machining the die and punch.

We start by briefly discussing the sheet metal part that is to be formed, as well as the tooling design and the sheet metal forming simulation. We then discuss the steps of adding a HAAS mill to CAMWorks in support of machining simulation and post processing for G-codes. We only discuss the virtual machining of the die to avoid repetition. At the end of the lesson, we briefly discuss the implementation of the forming process using the manufactured tooling for forming the part on the shop floor. The goal of the lesson is to provide readers a flavor of the practical aspect of virtual machining carried out in CAMWorks for support of a practical application. In addition, we discuss the detailed steps in adding a virtual HAAS mill and an associated post processor to CAMWorks. You may acquire computer files of virtual machines, similar to those to be discussed in Section 10.5, from your machine vendor, and follow the same steps in Section 10.5 to add a virtual machine to CAMWorks that replicates the physical machine at your shop floor.

The part example's files, including *part_1_draw_die_r1.SLDPRT* and *part_1_draw_die_r1 with toolpath.SLDPRT*, are available at the publisher's website for download. You may open the model file with toolpath; i.e., *part_1_draw_die_r1 with toolpath.SLDPRT*, to preview the toolpaths created for the die block. You may also open the model file, *part_1_draw_die_r1.SLDPRT*, and follow the discussion in Section 10.7 to create toolpaths and machining simulation similar to those discussed in Section 10.8. Instead of using a HAAS virtual machine, you may use your own virtual machine or the legacy machine, for example, *Mill_Tutorial*, that came with CAMWorks, to carry out machining simulation using the Machine Simulation capability of CAMWorks.

10.2   The Sheet Metal Part Design

The sheet metal part to be formed is a half clamp of a fuel line in an aerospace engineering system. The half clamp part shown in Figure 10.1(a) is symmetric in geometry, and is made of Alclad aluminum alloy 2024 with thickness 0.05 in. Key dimensions of the part are given in Figure 10.1(b). As can be expected that forming the part in one shot can be difficult due to the double curvature around the neck and the bends at ears, which induce both tearing and wrinkles during the forming process. Usually, such a part with delicate geometry requires multiple steps (e.g. bending the ears after the main body is formed). However, since one-shot forming saves both man hours and tooling cost, the focus was to develop a forming process and tooling that form the part to its desired shape through one single forming operation.

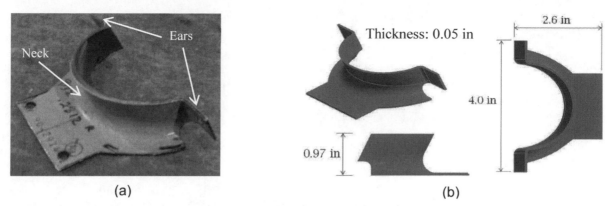

(a)

(b)

Figure 10.1 The half clamp part with double curvature (a) physical sample part, (b) CAD model with key dimensions.

10.3   Sheet Metal Forming Simulation

Forming simulations were first conducted to explore the formability of the part. The die face, including the surface of the two deep pockets in the die, was designed based on the geometric shape of the sheet metal part.

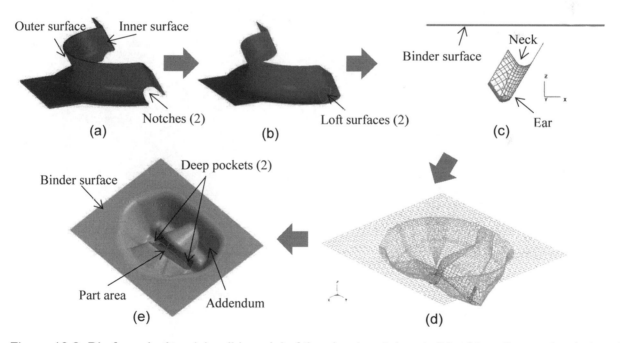

Figure 10.2 Die face design, (a) solid model of the sheet metal part, (b) mid-surface extracted and notches filled with loft surfaces, (c) flat binder surface created (side view), (d) addendum created, and (e) the complete die face.

The mid-surface of the part was first extracted as a surface model in SOLIDWORKS by compressing the outer and inner surfaces of the solid model; see Figure 10.2(a). The notches near the ears of the surface model were filled with loft surfaces; see Figure 10.2(b). The revised surface model was brought into DynaForm (www.dynaform.com) for die face design in support of forming simulations. The part was carefully oriented to prevent a potential undercut. In the meantime, a flat binder surface was created above the part; see Figure 10.2(c). An addendum surface that connects the binder surface and the part area was then automatically generated in DynaForm; see Figure 10.2(d). By trimming the binder surface with the addendum, the complete die face, shown in Figure 10.2(e), consisting of the part area, the addendum, and the trimmed binder surface, was obtained.

The die face was used as the female tool, while the male punch was modeled by copying and offsetting the geometry of the die face. Figure 10.3 illustrates a setup of draw forming simulation, in which the blank was pressed down by binders. Note that the geometric shape of the blank shown in Figure 10.3 was obtained after numerous unsuccessful blank designs, in which only half of the blank was modeled due to symmetry.

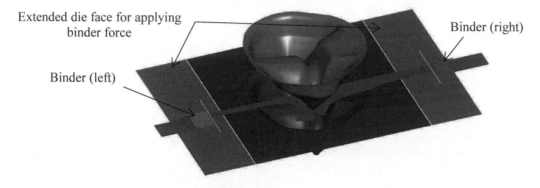

Figure 10.3 Draw forming simulation setup with binders

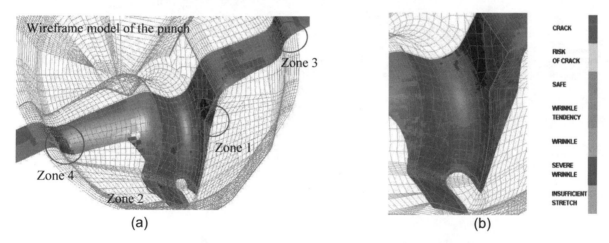

Figure 10.4 Draw forming simulations (with part boundary sketched in orange color in the half blank), (a) four zones with tearing (red color), and (b) successful forming simulations with modified die surface and blank shape (only mild wrinkles around the neck in pink color)

Initial forming simulations like that of Figure 10.4(a) reveal four areas with severe tearing. Tearing at Zones 3 and 4 is clearly due to the sharp transition in the die face geometry entering the cavity. The transition area of the die face was then modified to allow a smoother transition in geometry that facilitates material flow during forming. Tearing at Zones 1 and 2 is caused by the fact that the ears were bent early during the forming process and then dragged by the punch down to the deep pockets of the die, resulting in insufficient material flow from the right, and hence excessively stretching the blank material to the left of the ears. The blank shape was further modified aiming at allowing the ear to slide around the tip of the punch without being bent, which would contribute to a more evenly distributed material flow. This turned out to be a key factor that led to a successful forming of the part on the shop floor. Forming simulation of the modified tooling and blank is shown in Figure 10.4(b), in which tearing is completely eliminated and yet only mild wrinkling appears around the neck area. Such simulations show sufficient promise that merits moving forward to the next stage in the development, that is, the tooling design.

10.4   Tooling Design

The finalized die face geometry was exported from DynaForm and imported into SOLIDWORKS for tooling design. The tooling was designed to support a forming scenario, in which the blank is fixed at its left end and pressed at its right end by binder. To implement the scenario, the blank is extended in the length direction, and its left end is wrapped over the side face of the die block, as illustrated in Figure 10.5(a) and (b). The left end of the blank was designed to be bent three times and held tightly against the die block with two clamps and twelve bolts in order to avoid slippage during forming. Two bolt holes were drilled at the right side of the top face of the die block, as shown in Figure 10.5(c), so that the clamp could be installed on the top face of the die block to provide a holding force that could be adjusted by tightening up the two bolts at different torque levels. In addition, two shallow slots were cut at the edges of the block, as shown in Figure 10.5(d), to facilitate the transverse alignment of the blank.

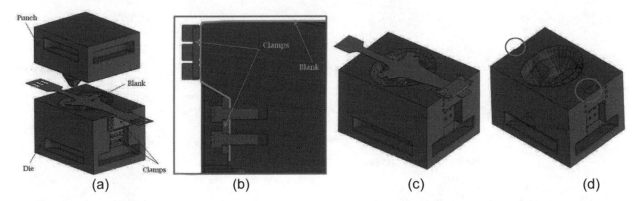

(a)   (b)   (c)   (d)

Figure 10.5 Tooling designed in SOLIDWORKS, (a) positioning of blank, punch, die, and clamps (b) the left side of the blank is bent and held by two clamps to mimic a fixed boundary condition, (c) the right side of the blank is held by a clamp on the top face of the die block to mimic a binder condition, and (d) shallow slots designed for transverse blank alignment

10.5   Adding a Virtual CNC Mill to CAMWorks

We have access to HAAS mills at our machine shop at the university. Therefore, we first added a virtual HAAS mill to CAMWorks to facilitate virtual machining simulation for tooling manufacturing. Adding a virtual mill to CAMWorks supports the following. First, it allows the user to use a virtual machine in CAMWorks that is a replica of the physical machine on the shop floor. Second, it offers users a customized post processor that converts toolpath to G-codes ready for machining without (or with minimum) modification. And third, it allows the virtual machine to be used in Machine Simulation like that of Lesson 7, in particular, detecting collision in machining operations.

Depending on the machine type there are at least three files needed. Files of extensions .ctl and .lng contain information on a customized post processor. The file with .kin extension defines the kinematics of the virtual machine. These files are added to the folder *C:CAMWorksData\CAMWorks2016X64\Posts*. If you list the contents of the folder, you should see several tutorial post processors came with CAMWorks, such as *M3AXIS-Tutorial*, *T2AXIS-Tutorial*, and *MT2AXIS-Tutorial*, that we used in the previous lessons.

We added three files to this folder: *Haas-5ax.ctl*, *Haas-5ax.kin*, and *Haas-5ax.lng*. You may do the same if you have acquired similar files from your machine vendors.

Note that in the folder, you may see files with extensions .pinf and .rtf. Files with .pinf extension are optional, which determines the default output file extension of the NC programs. The .rtf file is also optional, which displays information about the post processor when you select it in CAMWorks.

In addition, to support machine simulation, a set of geometric files that support the visualization of the virtual machine were added to the folder *C:CAMWorksData\ CAMWorks2016X64\MachSim\XML*. There are several subfolders under *XML* that came with CAMWorks, including *Mill_Tutorial* and *MillTurn_Tutorial*, which are generic machines supporting machine simulation. We chose one of them, *Mill_Tutorial*, for machine simulation in Lesson 7.

10.6   Customizing Technology Database

After adding files to CAMWorks data folders, we added the machine to CAMWorks by customizing its technology database, TechDB™. Here is what we did.

From the SOLIDWORKS pull-down menu, choose *Tools > CAMWorks > Technology Database*.

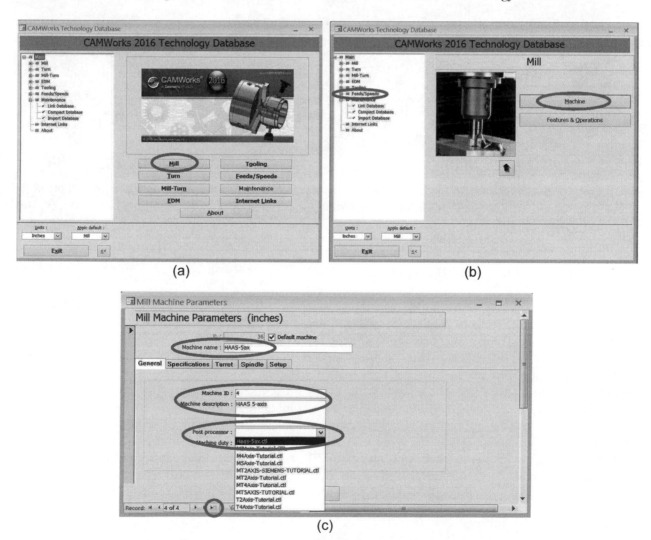

Figure 10.6 The *CAMWorks 2016 Technology Database* dialog box, (a) selecting *Mill* in the main window, (b) selecting *Machine*, and (c) entering machine information

The *CAMWorks 2016 Technology Database* dialog box appears; see Figure 10.6(a). Click *Mill*, as shown in Figure 10.6(a), and then click the *Machine* button; see Figure 10.6(b).

Click the *NEW (blank) record* button ⊞ —at the bottom, as circled in Figure 10.6(c)—to create a new machine (or a new record). Enter the machine name, for example: *HAAS-5ax*, Machine ID: *4*, and a short machine description. Then pull down the *Post processor* to choose *Haas-5ax.ctl* (which was added to the post processor folder at C:*CAMWorks data*).

You may click other tabs, such as *Specifications*, to add more specific information for the new mill being added to CAMWorks. Click the *Close* button to close the dialog box.

You may modify tools to match with those on the shop floor by expanding the *Tooling* entity and choosing, for example, *End Mills* [Figure 10.7(a)]. Four end mill types appear. Pick *Flat End Mill* to bring up the *Tool Database*, as shown in Figure 10.7(b), in which 745 flat-end mills available in the database are listed.

You may click the *NEW (blank) record* button ⊞ —at the bottom, as circled in Figure 10.7(b)—to create a new flat-end mill. Click the *Close* button to close the dialog box after adding new flat-end cutters.

You may modify feedrates and other machining parameters to reflect your practice on the shop floor. You may expand the *Feeds/Speeds* entity from the *CAMWorks 2016 Technology Database* window, which is circled in Figure 10.6(b), and choose, for example, *Feed/Speed Editor*. A *Micro Estimating Material Library* window appears, as shown in Figure 10.8(a), with *Material List* tab selected listing a total of 31 materials.

You may add a new material to the list by choosing *New* from the *File* pull-down menu.

Click the *Speeds & Feeds* tab to show machining parameters in the database; see Figure 10.8(b). You may modify the data by directly editing the number in a data cell.

You may continue editing the data in the database, or click the *Exit* button in the main window to exit the technology database.

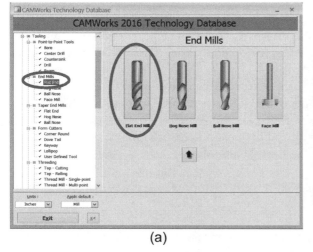

(a)   (b)

Figure 10.7 Adding end mills to CAMWorks technology database, (a) the *CAMWorks Technology Database* window, and (b) the *Tools Database – Mill* dialog box

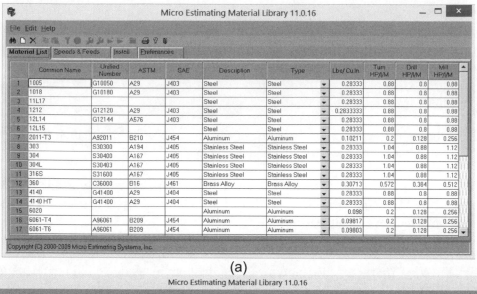

(a)

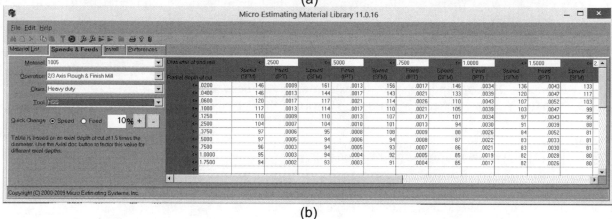

(b)

Figure 10.8 The *Micro Estimating Material Library* window, (a) materials listed under *Material List* tab, and (b) parameters under *Speeds & Feeds* tab

10.7 Virtual Machining Simulation using CAMWorks

In this section, we discuss the virtual machining of the die, in particular the machining operations that cut the cavity of the die [Figure 10.9(a)] from a 12in.×9in.×7.5in. Kirksite block, as shown in Figure 10.9(b). The material removal simulation is shown in Figure 10.9(c).

We used the HAAS-5ax mill added to CAMWorks and generated four operations: volume roughing, local milling, and two surface millings. Note that two surface milling operations are required to achieve desired surface finish of a maximum scallop 0.0003in. We also picked the post processor *Haas-5ax.ctl* to convert the toolpaths to G-codes compatible with the HAAS mill at the machine shop.

You may open the model file downloaded from the publisher's website to browse the toolpaths of the die machining. Again, the file name is *part_1_draw_die_r1 with toolpath.SLDPRT*.

The tools and key machining parameters employed for the four operations are listed in Table 10.1.

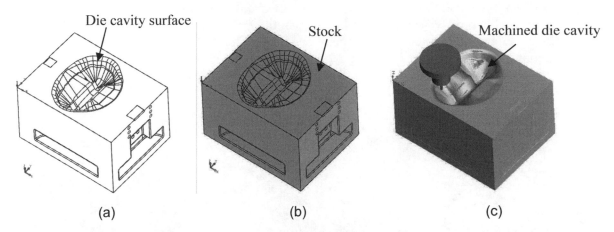

Figure 10.9 Virtual machining of the die block example, (a) the die design model with cavity surface, (b) stock, a rectangular Kirksite block, and (c) the material removal simulation of the die cavity machining operations

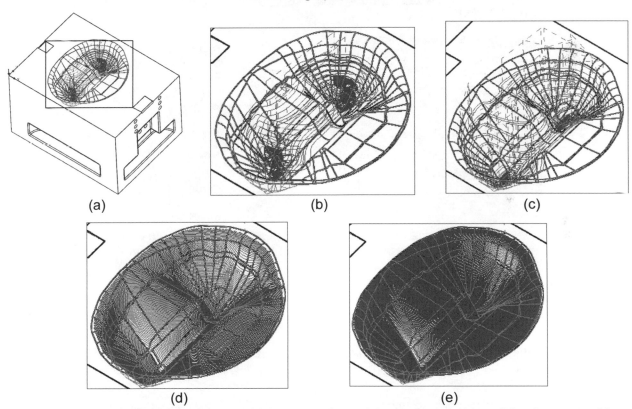

Figure 10.10 Toolpaths of the four machining operations, (a) volume roughing, (b) volume roughing, a closer view, (c) local milling, (d) surface milling 1, and (e) surface milling 2

Table 10.1 The tools and key machining parameters employed for the four operations

Operations	Tool	Stepover	Depth of cut	Scallop height
Volume Roughing	1 in. flat-end	0.25 in.	0.25 in.	
Local Milling	0.5 in. ball-nose	0.1 in.	0.1 in.	
Surface Milling 1	0.5 in. ball-nose			0.005 in.
Surface Milling 2	0.25 in. ball-nose			0.0003 in.

The following was what we did. We right clicked *Machine* and chose *Edit Definition*. In the *Machine* tab of the *Machine* dialog box, we selected *HAAS-5axis*. In the *Post Processor* tab, we chose *HAAS-5AX* for post processor. We then generated the four operations. The toolpaths and material removal simulations of the four opertations are shown in Figure 10.10 and Figure 10.11, respectively. Note that a 0.25in. ball-nose cutter is able to reach deep enough to the bottom of the two pockets, for final clean up and surface polishing, as can be seen by comparing Figure 10.11(d) and Figure 10.11(e).

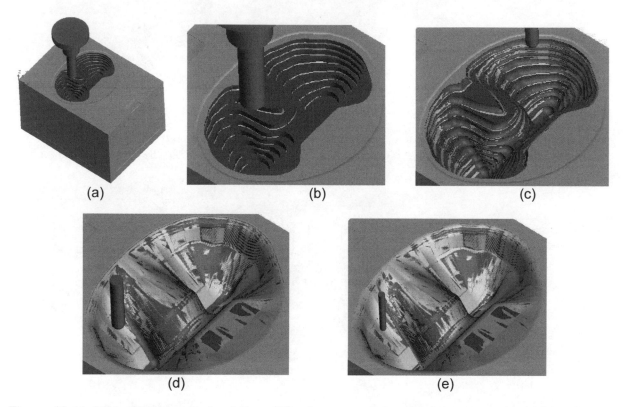

Figure 10.11 Material removal simulation of the four operations, (a) volume roughing, (b) volume roughing, a closer view, (c) local milling, (d) surface milling 1, and (e) surface milling 2

10.8   Machine Simulation

We chose the HAAS virtual machine added to CAMWorks to carry out the machine simulation.

Recall what we discussed in Lesson 7. To bring up the machine simulation, we right clicked *Mill Part Setup1* under CAMWorks operation tree 🅴 , and chose *Machine Simulation* and then *Legacy*.

The default *Mill_Tutorial* machine appeared like that of Lesson 7. We selected the HAAS mill for simulation by pulling down the *Machine* selection and choosing *5AxHaasVF3-XT-210* as the machine (see Figure 10.12).

Then we clicked the *Select* button ▢ next to the *Machine* on top of the *Machine Simulation* window (circled in Figure 10.13) to bring up the *Select Point* dialog box (Figure 10.14).

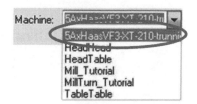

Figure 10.12 Choosing *HAAS mill* as the machine

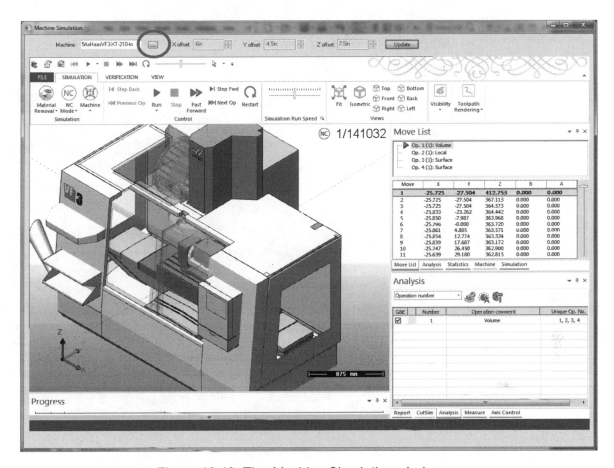

Figure 10.13 The *Machine Simulation* window

We picked the coordinate system, *Coordinate System1*, in the graphics area (see Figure 10.15), and entered 6in., 4.5in., and -7.5in. for the X-, Y-, Z-offsets, respectively, in the *Select Point* dialog box (Figure 10.14). These selections positioned the stock at the center of the rotary table mounted on the tilt table in the HAAS mill like that of Figure 10.16. We clicked the *Update* button on top; the HAAS mill appeared in the graphics area like that of Figure 10.13.

We clicked the *Machine Housing* button above the graphics area and chose *Hide* to hide the machine housing. We then rotated the view to see that the stock sits at the center of the top face of the rotary table, as shown in Figure 10.16.

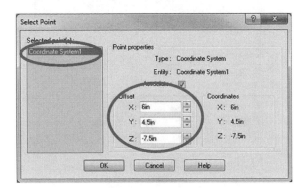

Figure 10.14 The *Select Point* dialog box

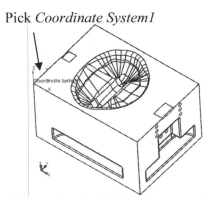

Pick *Coordinate System1*

Figure 10.15 Picking the coordinate system, *Coordinate System1*

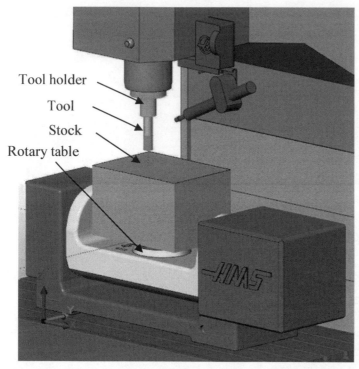

Figure 10.16 The stock sitting at the center of the top face of the rotary table

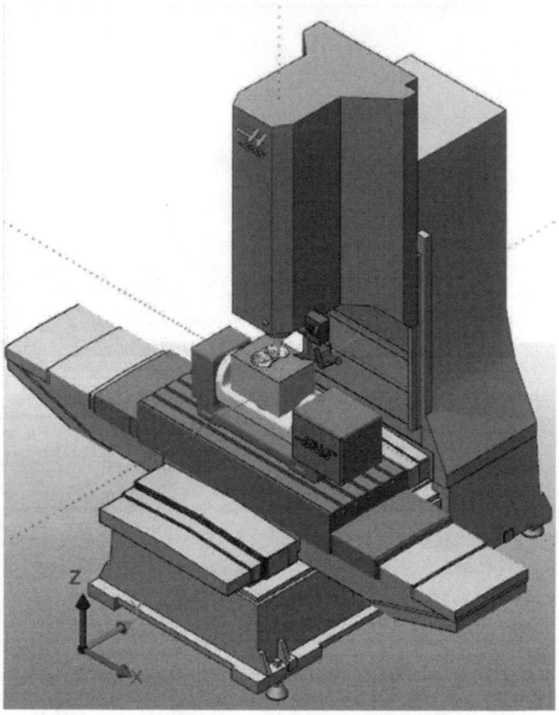

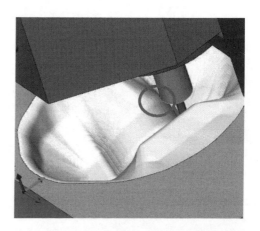

Figure 10.17 Running simulation for all four operations

We clicked the *Run* button ▶ to start the simulation. The G-codes of the corresponding tool motion were displayed in the lower portion of the *Move List* area.

The simulation ran to the end for the first three operations without any issue (Figure 10.17). However, collisions were detected between the tool holder and the die surface when the 0.25in. ball-nose cutter was reaching the bottom of the pockets in the final operation, *Surface Milling 2*, as circled in Figure 10.18. To avoid the collisions, a longer cutter (4.5in. in length) and a tool holder of smaller diameter (0.75in.) were employed.

The same cutter and holder were used when we physically cut the die on the shop floor.

We followed the same steps as discussed in Lesson 2 to convert the toolpaths into G-codes by choosing the post processor *HAAS-5AX* (right click *Machine*, choose *Edit*, and click the *Post Process* tab).

Figure 10.18 Collision encountered between the tool holder and the die surface

10.9 Machining the Die Using HAAS Mill

In addition to the toolpath that cuts the die cavity as discussed above, other toolpaths were generated to machine the notches on the top face, slots on the four side faces, and tap the holes. Similar toolpaths were created to machine the punch as well.

The G-codes converted were then uploaded to a HAAS mill to machine the die (and punch, see Figure 10.19). The machined punch and die (cut out from two Kirksite blocks) as well as one of the aluminum clamps are shown in Figure 10.20(a). The machined tooling blocks, including the die, punch, and clamps, are of excellent quality with desired surface finish. The tooling is ready for forming the part. Figure 10.20(b) shows the implementation of the scenario discussed in Section 10.3.

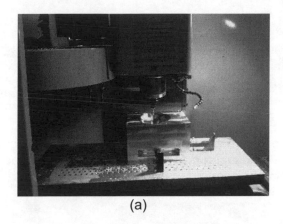
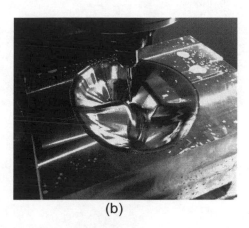

(a)   (b)

Figure 10.19 Machining the die on a HAAS mill, (a) the Kirksite block mounted on the jig table, and (b) a closer look at the machined die block

10.10 Process Implementation and Validation

Forming attempts were conducted on a 300 ton press on the shop floor. Tearing was found in all attempts near Zone 1 shown in Figure 10.4(a), indicating that the material flow was not sufficient. After the first failed attempts, a few more blanks were formed by adding a lubricant between the tooling and the blank to increase material flow, and at the same time lowering the blank holding force to minimum (bolts finger tighten). Although with the increasing material flow, smaller cracks still appeared on the blank at about the same location.

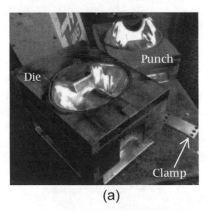

(a)   (b)

Figure 10.20 Manufactured tooling blocks, (a) punch, die and clamps machined, and (b) implementation of the forming scenario with blank fixed at its left side and clamped on its right side

As a last resort, two forming tricks were implemented based on the finding in simulation that the ears on the blank cannot be bent and dragged into the deep pockets, which was believed to be the major cause of tearing on the failed blanks. By bending the ears beforehand, the contact area between the ears and the punch was significantly reduced, which helped the ears to slide away from the tip of the punch, thus avoiding being dragged into the deep pockets. In addition, to maximize the material flow from the right, the forming process was paused at about 0.75in. before closure between the punch and die. The partially formed blank was taken out of the tooling and trimmed off its right portion to completely remove holding force from the right. The trimmed blank was then brought back to the tooling and the forming was resumed until the punch and die closed completely. This process is illustrated in Figure 10.21.

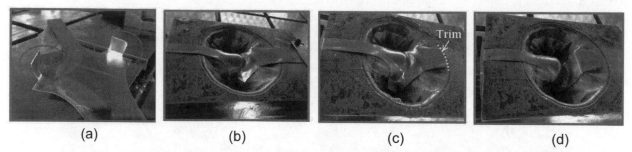

     (a)   (b)   (c)   (d)

Figure 10.21 Forming tricks implemented on the shop floor: (a) bending the ears in advance, (b) forming the blank until the punch is 0.75 in. before closing, (c) trimming off the right side of the blank to allow more material flow, and (d) continuing forming until die and punch close.

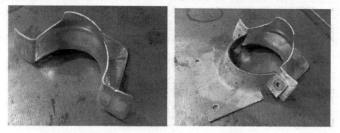

Figure 10.22 Successfully formed part (left) with the matching half in yellow color (right).

With the implementation of the two tricks, the ears were able to slide away without being dragged into the deep pockets. A blank was formed at the end with no tearing or wrinkle, producing a working part that fit well with the other half of the fuel line clamp, as shown in Figure 10.22. The same forming process was repeated several times and the same working parts were formed successfully, which verified the repetitivity of the process.

The successfully formed part was scanned into the computer using ATOS III scanner (www.gom.com). The scanned geometry was compared with the CAD solid model of the part design for accuracy verification. As shown in Figure 10.23, the deviation between the original CAD design and the scanned part was between ±0.04 in, which satisfied the engineering requirements for the part. The project came to a perfect closure at this point.

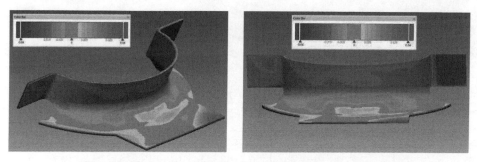

Figure 10.23 Deviation check for the formed part.

Appendix A: Machinable Features

A.1 Overview

Feature-based machining is the driving concept behind CAMWorks. By defining areas to be machined as features, CAMWorks is able to apply more automation and intelligence into toolpath creation. In this appendix, we discuss machinable features that are extracted by using feature recognition capabilities provided by CAMWorks, including automated feature recognition (AFR), interactive feature recognition (IFR) and local feature recognition (LFR). We focus more on machinable features for milling operations, including the NC operations generated to the machinable features by the rules, called strategy, implemented in CAMWorks technology database. We also briefly discuss machinable features for turning operations.

A.2 Machinable Features for Milling Operations

Toolpaths can be generated only on machinable features. CAMWorks provides three methods, AFR, IFR, and LFR, for defining machinable features for milling.

Automatic Feature Recognition (AFR)

Automatic Feature Recognition analyzes the part geometric shape and attempts to extract most common machinable features such as pockets, holes, slots and bosses. You may select the *Extract Machinable Features* button 🔳 above the graphics area, choose from the pull-down menu *Tools > CAMWorks > Extract Machinable Features*, or right click *Mill Part Setup* (under CAMWorks feature tree 🔳) and choose *Recognize Features* to initiate AFR. Depending on the complexity of the part, AFR can save considerable time in defining 2.5 axis features, including prismatic features commonly found in milling operations.

Most 2.5 axis features can be extracted automatically. One of the major characteristics of the 2.5 axis features is that the top and bottom of the feature are flat and normal to the tool axis of the machining operations, including prismatic solid feature and solid features with tapered wall. Typical 2.5 axis features can be a boss, pocket, open pocket, corner slot, slot, hole, face feature, open profile, curve or engrave feature. Some of these features are illustrated in Figure A.1.

The types of 2.5 axis features to be included in AFR can be selected in CAMWorks. You may choose from the pull-down menu *Tools > CAMWorks > Options* to bring up the *Options* dialog box like that of Figure A.2.

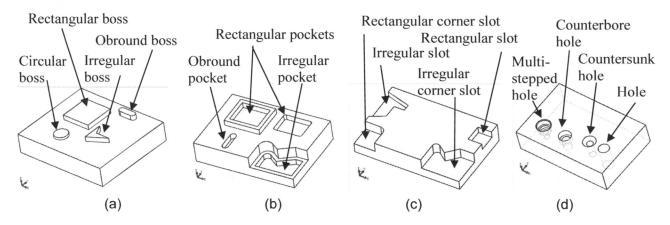

Figure A.1 Illustration of 2.5 axis features, (a) bosses, (b) pockets, (c) slots, and (d) holes

If you click the *Mill Features* tab of the *Options* dialog box, the default feature types to be extracted are listed. By default, holes, non holes (such as pocket, slot, etc.), boss, and tapered & filleted are selected.

Also, *MfgView* is selected for *Method*. Two methods are currently provided: *AFR* and *MfgView*.

When *AFR* is selected, CAMWorks analyzes the SOLIDWORKS solid model and identifies two-dimensional prismatic and tapered wall machinable features. When *MfgView* is selected, CAMWorks uses an alternative method to generate features and finds additional feature types not found by *AFR*.

Figure A.2   The *Options* dialog box

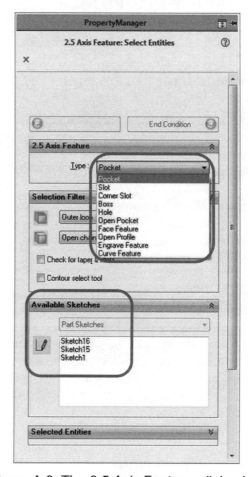

Figure A.3   The *2.5 Axis Features* dialog box

The following feature types are currently supported in CAMWorks:

- Bosses and pockets with vertical walls.
- Bosses and pockets with constant tapered walls with or without constant radius bottom or top fillets and chamfers.
- Pockets and bosses are further broken down into rectangular, circular, obround and irregular.
- Slots with vertical walls.
- Slots have categories for rectangular, corner rectangular, irregular, and corner irregular.
- Numerous hole types including simple holes, counterbores, countersinks, and multi-stepped holes.
- Simple holes can also be described as being drilled, bored, reamed, or threaded.

Local Feature Recognition (LFR)

CAMWorks provides a selective form of Automatic Feature Recognition (AFR) based on user-selected faces in the form of Local Feature Recognition (LFR). This is a semi-automatic method to define features based on face selection. Single or multiple features can be extracted depending on the selected faces.

You may pick one or more faces of the part on which local features are to be recognized. Then, right-click *Mill Part Setup* under the CAMWorks feature tree tab [CW] and select *Recognize Local Features* from the context menu. After executing this command, the locally recognized features, if any, will be added to the machinable feature list.

Interactive Feature Recognition (IFR)

Interactive Feature Recognition allows you to extract interactively either 2.5 axis or multi surface features.

As you may be aware, AFR cannot recognize every feature on complex parts and does not recognize some types of features. To machine these areas, you need to define features interactively by right clicking *Mill Part Setup* (under CAMWorks feature tree tab [CW]) and choosing, for example, *New 2.5 Axis Feature*. In the *2.5 Axis Features* dialog box (Figure A.3), select feature type and pick a sketch of the feature to create a machinable feature interactively. Detailed steps for creating such a machinable feature can be found in the book, for example, Lesson 2: Simple Plate.

If you activate CAMWorks *3 Axis Milling* (as discussed in Lesson 1, see Figure 1.14), you may define multi surface features interactively by specifying faces to be cut and faces to avoid. Detailed steps for creating such machinable and avoid features can be found in Lesson 4: Machining a Freeform Surface.

As discussed in Lesson 1, CAMWorks uses a set of knowledge-based rules to assign machining operations to machinable features. The technology database (or TechDB™) contains data and rules that determine operation plans for the respective machinable features. These data and rules can be customized to meet your specific needs. The allowable milling operations for the respective machinable features established in TechDB™ are summarized in Table A.1.

Table A.1 A list of allowable milling operations for the respective machinable features

Machinable Feature	Description	Allowable Operations
Rectangular Pocket	A pocket whose general shape is a rectangle. The corners of the rectangle can be either sharp or radiused.	Rough Mill, Contour Mill
Circular Pocket	A pocket whose shape is defined by a circle. The difference between a circular pocket and a hole is that a circular pocket can contain an island.	Rough Mill, Contour Mill, Drill, Bore, Ream, Tap, Center Drill, Countersink
Irregular Pocket	A pocket whose shape is neither rectangular nor circular.	Rough Mill, Contour Mill
Obround Pocket	A rectangular shaped boss with 180 degree round ends.	Rough Mill, Contour Mill
Rectangular Slot	A pocket with one edge open to the outside of the part. Machining is extended beyond the open segment of the slot. The general shape of the feature is rectangular. The corners of the rectangle can be either	Rough Mill, Contour Mill

	sharp or radiused.	
Irregular Slot	Similar in definition to a Rectangular Slot except that the general shape is not rectangular.	Rough Mill, Contour Mill
Rectangular Corner Slot	Similar to a Slot except that the feature can contain two or more adjacent edges that are open to the outside of the part. Machining is extended beyond each of these open edges. The corners of the rectangle can be either sharp or radiused.	Rough Mill, Contour Mill
Irregular Corner Slot	Similar to a Rectangular Corner Slot except that the general shape is not rectangular.	Rough Mill, Contour Mill
Rectangular Boss	A Boss whose general shape is rectangular. The corners of the rectangle can be either sharp or radiused.	Contour Mill
Circular Boss	A Boss whose shape is round.	Contour Mill, Thread Mill
Irregular Boss	Any Boss whose shape is neither rectangular nor circular.	Contour Mill
Obround Boss	A rectangular shaped boss with 180 degree round ends.	Contour Mill
Hole	Same shape as a SOLIDWORKS Solids Simple hole with a blind or through end condition or a Simple Drilled hole. Strategies can be assigned as Drill, Bore, Ream, or Thread.	Rough Mill, Contour Mill, Drill, Bore, Ream, Tap, Center Drill, Countersink, Thread Mill
Countersunk Hole	Same shape as a SOLIDWORKS Solids Countersunk hole with a blind or through end condition or a C-Sunk Drilled hole. Strategies can be assigned as Drill, Bore, or Ream.	Rough Mill, Contour Mill, Drill, Bore, Ream, Tap, Center Drill, Countersink, Thread Mill
Counterbored Hole	Same shape as a SOLIDWORKS Solids Counterbored hole with a blind or through end condition or a C-Bored Drilled hole. Strategies can be assigned as Drill, Bore, or Ream.	Rough Mill, Contour Mill, Drill, Bore, Ream, Tap, Center Drill, Countersink, Thread Mill
Countersink/Counterbore Combination	Each countersink/counterbore combination will be recognized as 2 separate hole features that are based on the Mill Part Setup direction.	Rough Mill, Contour Mill, Drill, Bore, Ream, Tap, Center Drill, Countersink, Thread Mill
Multi-stepped Hole	Any hole with multiple steps that is not recognized as a hole, countersunk hole or counterbored hole.	Rough Mill, Contour Mill, Drill, Bore, Ream, Tap, Center Drill, Countersink, Thread Mill
Open Pocket	Created only when AFR finds a Boss feature and the machining direction is parallel to one side of the stock. The bottom of the open pocket is the bottom of the boss. The boss becomes an island in the open pocket.	Rough Mill, Contour Mill
Face Feature	If the Face option is checked on the Features tab in the Options dialog box, a Face Feature is created when AFR finds at least one non-face machinable feature, the topmost face is parallel to the Mill	Face Mill, Rough Mill, Contour Mill

	Part Setup and the machining direction is parallel to one of the sides of the stock.	
Perimeter–Open Pocket or Perimeter-Boss Feature	If the Perimeter option is checked on the Features tab in the Options dialog box, AFR creates either a boss or open pocket feature for the perimeter for any Setup created via AFR.	Rough Mill, Contour Mill, Face Mill

A.3 Machinable Features for Turning Operations

Similar to milling operations, machinable features can be extracted in CAMWorks either automatically using Automatic Feature Recognition (AFR) or created interactively by defining a turn feature from a SOLIDWORKS part face, sketch or edge.

The following turn feature types are currently supported:

- Face Feature: A Face feature is defined from vertical edges at the front edge of the part model.
- OD Feature: An OD feature includes the outside shape of the part from the face feature to the cutoff feature, not including the shape of any groove features.
- ID Feature: An ID feature includes the inside diameter shape of the part from the Face feature to the Cut Off feature, not including the shape of any groove features.
- Groove Feature with vertical walls: A Groove is a feature that is closed on both sides and below the surface of the surrounding geometry. There are three categories of grooves: rectangular, half obround and generic. You can cut a groove into the outer diameter, the inner diameter or a face. Groove features where the bottom of the groove is not parallel to the Z or X axis are not recognized by AFR and must be defined interactively.
- Cut Off Feature: A Cut Off feature is defined from vertical edges on the opposite side of the Face feature. A Cut Off feature is similar to a face and can be converted to a Face feature for two-step turning operations using the Convert to Face Feature command on the Cut Off feature context menu.

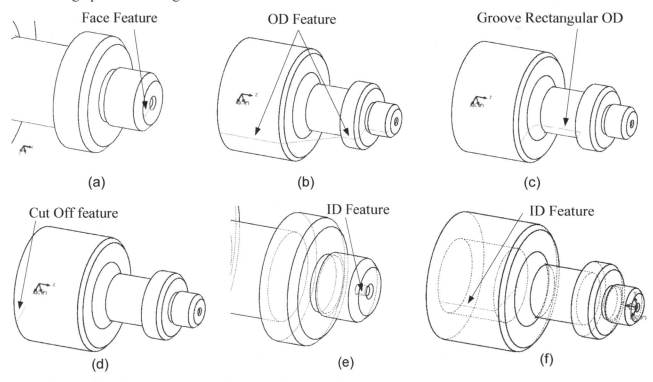

Figure A.4 Examples of machinable features of turning operations, (a) *Face Feature*, (b) *OD Feature*, (c) *Groove Rectangular OD*, (d) *Cut Off Feature*, (e) *ID Feature* (front end hole), and (f) *ID Feature* (rear end hole)

Some of the aforementioned features are illustrated in Figure A.4. The allowable turning operations for the respective machinable features established in TechDB™ are summarized in Table A.2.

Table A.2 A list of allowable turning operations for the respective machinable features

Machinable Feature	Description	Allowable Operations
Face	A Face feature is defined from vertical edges at the front of the part model. This feature is generally used to trim excess stock off the front of the part.	Face Rough Face Finish
OD	An OD (outer diameter) feature includes the outside shape of the part from the face feature to the cutoff feature, excluding the shape of any groove features. Typically, this feature is used to trim the stock away, leaving you with the outer shape of your design.	Turn Rough Turn Finish Face Rough Face Finish Groove Rough Groove Finish Thread
ID	An ID (inner diameter) feature includes the inside diameter shape of the part from the face feature to the cutoff feature, excluding the shape of any groove features.	Bore Rough Bore Finish Drill Thread
Groove Rectangular OD	A Groove Rectangular OD is a recessed feature on the OD of the part where the side walls are equal and parallel to the X machining direction. The bottom and or top may include constant radius fillets or equal size chamfers.	Groove Rough Groove Finish Turn Rough Turn Finish
Groove Rectangular ID	A Groove Rectangular ID is a recessed feature on the ID of the part where the side walls are equal and parallel to the X machining direction. The bottom and or top may include constant radius fillets or equal size chamfers.	Groove Rough Groove Finish Bore Rough Bore Finish
Groove Rectangular Face	A Groove Rectangular Face is a recessed feature on the front edge of the part where the side walls are equal and parallel to the Z machining direction. The bottom and or top may include constant radius fillets or equal size chamfers.	Groove Rough Groove Finish
Groove Half Obround OD	A Groove Half Obround OD is a recessed feature on the OD of the part where the side walls are equal and parallel to the X machining direction. The side walls of the groove are joined at the bottom of the groove by a single 180 degree radius.	Groove Rough Groove Finish
Groove Half Obround ID	A Groove Half Obround ID is a recessed feature on the ID of the part where the side walls are equal and parallel to the X machining direction. The side walls of the groove are joined at the bottom of the groove by a single 180 degree radius.	Groove Rough Groove Finish
Groove Half Obround Face	A Groove Half Obround Face is a recessed feature on the Face of the part where the side walls are equal and parallel to the Z machining direction. The side walls of the groove are joined at the bottom of the groove by a single 180 degree radius.	Groove Rough Groove Finish
Groove Generic OD	A Groove Generic OD is any recessed groove shaped feature on the OD of the part that is neither rectangular nor half obround in shape.	Groove Rough Groove Finish Turn Rough Turn Finish
Groove Generic ID	A Groove Generic ID is any recessed groove shaped feature on the ID of the part that is neither rectangular nor half obround in shape.	Groove Rough Groove Finish Bore Rough Bore Finish

Groove Generic Face	A Groove Generic Face is any recessed groove shaped feature on the face of the part that is neither rectangular nor half obround in shape.	Groove Rough Groove Finish
Cut Off	A Cut Off feature is defined from vertical edges on the opposite side of a Face feature. This feature is used primarily to trim excess stock from the back of the part. A Cut Off feature is similar to a face and can be converted to a Face feature for two-step turning operations using the *Convert to Face Feature* command on the Cut Off feature context menu.	Cut Off Face Rough Face Finish Groove Rough Groove Finish

Appendix B: Machining Operations

B.1   Overview

When a machinable feature is extracted or created, the corresponding machining operations are generated by the knowledge rules and data stored in the TechDB™. In this appendix, we discuss briefly the operations involved in both milling and turning operations.

B.2   Milling Operations

Milling operations are grouped by the types of machinable features, i.e., 2.5 axis features, 3 axis cutting cycles, and multiaxis.

2.5 Axis Features

The operations that support machining 2.5 axis features include Rough Mill, Contour Mill, Face Mill, Thread Mill, and Single Point.

A Rough Mill operation removes material from a part by following the shape of the machinable feature (pocket-in, pocket-out, spiral in or spiral out) or by making a series of parallel cuts across the machinable feature (zigzag, zig or plunge rough).

A Contour Mill operation removes material from a part by following the shape of the profile of pockets, slots, bosses, etc.

The Face Mill operation generates a toolpath on a mill Face feature for squaring or facing off the top of a part. Although a Rough Mill operation can generate toolpaths that will remove material from a Face feature, the Face Mill operation provides specific controls to produce a toolpath motion that is more appropriate for this task.

Thread Mill operations create thread mill toolpaths for a hole or circular boss. A thread mill operation can be generated automatically by assigning a thread mill Strategy to a hole or circular boss and selecting Generate Operation Plan. Alternatively, you can insert a thread mill operation using the New Operation and New Hole Operation commands.

Single point cycles include Drill, Bore, Ream, Tap, Countersink, and Centerdrill.

3-Axis Cutting Cycles

The 3 Axis cutting cycles include Area Clearance, Pattern Project, Z Level, Constant Stepover, Pencil Mill, Curve Project, and Flat Area.

The Area Clearance cycle removes the material between the stock or contain area and the selected feature at decreasing Z depth levels by making a series of parallel cuts across the stock (Lace) or by pocketing out toward the stock.

The Pattern Project operation is a multi-surface finishing cycle that removes material based on the selected pattern: Slice, Flowline, Radial and Spiral. These patterns have unique characteristics that make them appropriate for semi-finishing and finishing selected areas or the entire model.

The Z Level cycle is a finish contouring cycle that removes material by making a series of horizontal planar cuts. The cuts follow the contour of the feature at decreasing Z levels based on the Surface Finish you specify. Cutting starts from the highest location on the model and works down.

The Constant Stepover operation removes material by maintaining a constant user-defined stepover relative to the surface.

The Pencil Mill cycle generates toolpaths to finish machine corner areas using a single pass or multiple constant stepover passes. Corner areas are defined where the radius of curvature on the feature is less than the radius of the tool.

The Curve Project cycle removes material by projecting selected 2.5 Axis Engrave and/or Curve features onto the faces/surfaces of a Multi Surface feature and generating toolpaths along the projected entities. CAMWorks can calculate a single pass or multiple depth passes for engraving.

The Flat Area cycle uses a pocket out pattern to remove material from feature faces that are flat and parallel to the XY machining plane. CAMWorks generates toolpaths only on completely flat areas. If a face/surface has even a small gradient, CAMWorks will not generate a toolpath. This cycle can be used for finishing where excess material has already been cleared and supports single or multiple depths of cut.

Multiaxis Operations

Multiaxis operations are generated to machine a single freeform or multiple surface features to achieve desirable surface finish.

B.3 Turning Operations

The Turn cutting cycles include Face Rough, Face Finish, Turn Rough, Turn Finish, Groove Rough, Groove Finish, Bore Rough, Bore Finish, Drill, Center Drill, Thread, and Cut Off.

Face Rough: The Face Rough Operation defines multiple rough cuts to machine turned faces.

Face Finish: The Face Finish Operation defines a single finish cut to machine turned faces.

Turn Rough: The Turn Rough operation defines multiple rough cuts to machine turned faces. This is a common machining operation for a part that has a range of stock on the ODs that requires several cuts to remove the bulk of the stock.

Turn Finish: The Turn Finish operation defines multiple rough cuts to machine turned faces.

Groove Rough/Finish: The Grooving Cycle (Rough and Finish) allows you to cut a groove of almost any shape.

Bore Rough/Finish: The Bore Cycle (Rough and Finish) allows you to bore a hole or an ID feature.

Drill: The Drill operation generates a drill toolpath at the center line of the part (Z axis).

Center Drill: The Center Drill operation generates a center drill toolpath at the center point of the hole with a small depth.

Thread: In CAMWorks Turning, you use Strategies to define a thread on a feature. In order to generate a Thread operation, the corresponding thread condition must be selected from the TechDB™. The major diameter for the thread condition in the TechDB™ must match the feature's maximum diameter.

Cut off: When the stock is defined as bar stock, Automatic Feature Recognition generates a Cut off feature on the opposite side of the Face feature. A Cut off operation is generated for a Cut off feature when you generate an Operation Plan.